The Genial Gene

The Genial Gene

Deconstructing Darwinian Selfishness

JOAN ROUGHGARDEN

UNIVERSITY OF CALIFORNIA PRESS
Berkeley   Los Angeles   London

The publisher gratefully acknowledges the generous support of the General Endowment Fund of the University of California Press Foundation.

University of California Press, one of the most distinguished university presses in the United States, enriches lives around the world by advancing scholarship in the humanities, social sciences, and natural sciences. Its activities are supported by the UC Press Foundation and by philanthropic contributions from individuals and institutions. For more information, visit www.ucpress.edu.

University of California Press
Berkeley and Los Angeles, California

University of California Press, Ltd.
London, England

Library of Congress Cataloging-in-Publication Data

Roughgarden, Joan.
  The genial gene : deconstructing Darwinian selfishness / Joan Roughgarden.
    p. cm.
  Includes bibliographical references and index.
  ISBN 978-0-520-25826-6 (case : alk. paper) 1. Sexual selection in animals. 2. Sexual behavior in animals. 3. Reproduction. I. Title.
  QL761.R68 2009
  591.56'2—dc22

                                                          2008048660

Manufactured in the United States of America

16 15 14 13 12 11 10 09
10 9 8 7 6 5 4 3 2 1

The paper used in this publication meets the minimum requirements of ANSI/NISO Z39.48-1992 (R 1997) (*Performance of Paper*).

To those who suffer
the persecution of science

Contents

Acknowledgments

I thank the members of my laboratory group for participating in the literature searches and discussion that have provided the foundation for this book, especially, Erol Akçay, Timothy Bonebrake, Kathleen Fitzgerald, Henri Folse, Priya Iyer, and Jeremy Van Cleve. I thank Deborah Gordon and Angela Potochnik for their encouragement and comments on the manuscript. I thank Jonathan Kaplan and Michael Wade for helpful reviews, and the staff of the University of California Press, especially my editor, Blake Edgar, for his interest and support of the project.

INTRODUCTION Is Nature Selfish?

This book is about whether selfishness and individuality, rather than kindness and cooperation, are basic to biological nature. Darwinism has come to be identified with selfishness and individuality. I criticize this evolutionary perspective by showing it misrepresents the facts of life as we now know them. I focus on social behavior related to sex, gender, and family where the reality of universal selfishness and sexual conflict is supposedly most evident. I show that writings in the professional biological literature advocating a picture of universal selfishness, as well as similar writings in books and articles aimed at the general public, are mistaken. I present my laboratory's alternative evolutionary theories for social behavior that emphasize cooperation and teamwork and that rely on the mathematics of cooperative games.

1

In a previous book, *Evolution's Rainbow*,[1] I offered a survey of diversity in gender and sexuality focusing on animals, together with a brief mention of humans across cultures and history. As I was writing *Evolution's Rainbow*, I became increasingly critical of how this diversity is ignored in biology curricula worldwide and critical of the language and theories purporting to describe and explain this diversity. The theory to account for sexual behavior in evolutionary biology is called "sexual selection," a topic that originates with Darwin's writings in 1871.[2] I concluded that Darwin's sexual-selection theory was completely false and needed to be replaced by some new and equally expansive theoretical system. I termed the replacement theory *social selection* but did little more than sketch a few points that such a new theoretical system might contain. I criticized the established theory and promised a new one someday.

Since 2004, my students and I have been publishing papers that develop mathematical models for issues such as why sex evolved, why the male/female binary evolved at both the gametic and whole-organism levels, whether a male and female are necessarily in conflict, and why males and females have offspring outside their pair bonds and raise young they have not parented. All these issues are central to understanding the reproductive social life of animals.

The topics of diversity in gender and sexuality that figured so prominently in *Evolution's Rainbow* receive little attention here, because a theory for such diversity depends on getting the foundational theory correct to begin with. Instead, the issue we face is whether sexual-selection theory offers a correct starting point for understanding *anything* having to do with reproductive social behavior, anything from peacock tails, to the worms robins feed their nestlings, to homosexuality. I don't think so.

Important and interesting as sexual selection may be in its own right, a still greater concern is the overall message that evolutionary biology imparts to the world. This book's title, *The Genial Gene*, alludes to the

---

1. Joan Roughgarden, *Evolution's Rainbow: Diversity, Gender and Sexuality in Nature and People* (Berkeley: University of California Press, 2004).

2. Charles Darwin, *The Descent of Man, and Selection in Relation to Sex* (Princeton: Princeton University Press, 1871 [facsimile ed]).

book, *The Selfish Gene*, written by Richard Dawkins in 1976[3] that extended and brought to general attention G. C. Williams' critique of group selection.[4] I remember teaching from *The Selfish Gene* when it first appeared and noticed the appeal that a naturalized doctrine of selfishness has to certain students and to those in the general public who, for example, identify with Ayn Rand's writings that celebrate the ethic of individualism.[5] This selfishness is now thought to be "nature's way" and evolutionary biology is thought to authorize the truth of selfish behavior throughout the living world. Evolutionary biology presently explains what little kindness and cooperation it wishes to acknowledge using work-around theories called *kin selection*[6] and *reciprocal altruism*.[7] These are theories whose purpose, as a proponent once told me, is to take the altruism out of altruism—theories that devise a way to see how cooperative behavior is really deep-down selfishness after all.

The selfish-gene philosophy is what I term *neo-Spencerism*. It claims to represent neo-Darwinism, which is what Darwinism has been called since genes were added to Darwin's original theory. However, the writings from the founders of neo-Darwinism, such as mathematical geneticists R. A. Fisher,[8] J. B. S. Haldane,[9] and S. E. Wright[10] in the 1930s and later, are relatively free of ideology. Similarly, Darwin's writings are relatively free of ideology compared to those of Herbert Spencer who coined the phrase, "survival of the fittest,"[11] which has become so emblematic of

3. Richard Dawkins, *The Selfish Gene* (Oxford: Oxford University Press, 1976).

4. George C. Williams, *Adaptation and Natural Selection*. Princeton: Princeton University Press, 1966.

5. Ayn Rand, *Atlas Shrugged* (Signet, 1957); Ayn Rand, *The Virtue of Selfishness: A New Concept of Egoism* (Signet, 1961).

6. Kin selection refers to an animal increasing its genes in the next generation by helping a close relative.

7. Reciprocal altruism refers to one animal helping another in expectation of the action being reciprocated some time later.

8. Ronald A. Fisher, *The Genetical Theory of Natural Selection* (Oxford: Oxford University Press, 1930).

9. J. B. S. Haldane, *The Causes of Evolution* (London: Longmans, Green and Co., 1932).

10. Sewall Wright, *Evolution and the Genetics of Populations*, vol. II of *The Theory of Gene Frequencies* (Chicago: University of Chicago Press, 1969).

11. Herbert Spencer, *The Principles of Biology*, vol. 1 (London: Williams and Norgate, 1864), 444.

Darwinism. The neo-Spencerists publicize "sexual conflict" as extending the selfish-gene metaphor into reproductive social behavior, including the "natural" roles for each gender and sex. Just as Spencer exaggerated and politicized the writings of Darwin, neo-Spencerists exaggerate and politicize the writing of neo-Darwinism's founders. They push selfishness and sexual conflict as the social message of neo-Darwinism just as Spencer pushed survival of the fittest as the social message of Darwinism.

Thus, the challenge I pose to sexual selection theory with its present-day embrace of the selfish gene and sexual conflict is important beyond understanding how many worms each bird brings to its nestlings and why the peacock has a colorful tail. What is being challenged is the scientific validity of a world view that naturalizes selfishness and sexual conflict.

When you hear two birds call in the morning, do you think they are lying to one another, planning how to cheat and steal from each other? Or could you think they may be coordinating their activities for the day's labor? The neo-Spencerists think the matter is already settled—birds, like the rest of the creatures in nature, supposedly spend their days lying, cheating, and stealing from one another.

In contrast, I think the matter is far from settled. Neo-Spencerists have not scientifically demonstrated their world view of nature. They have merely stipulated it and ridicule any alternative view of nature as romantic wishful thinking. Well, I challenge the neo-Spencerists to be the scientists they claim to be, to engage and entertain alternative hypotheses objectively without descending to the personal and homophobic rebukes that have characterized their discourse so far.[12]

The sexual-selection theory underwrites "genetic classism" by naturalizing a mythical urge on the part of females to locate and sleep with males who have the best genes. Sexual selection is a narrative of genetic entitlement. I think the practicality of an egalitarian society

---

12. Cf. documentation in: Joan Roughgarden, "Challenging Darwin's Theory of Sexual Selection," *Daedalus* 136 (2):1–14 (2007); UK homophobic backlash in: C. McKechnie and D. Shuker, *Journal of Literature and Science* 1 (1):75–76 (2007); and recent governmental summaries of UK homophobia in: www.pfc.org.uk/files/reporting_homophobia_in_the_health_sector.pdf and www.pfc.org.uk/files/ eurostudy.pdf.

depends on whether a rational rejection exists for sexual selection. If sexual selection is indeed true, then so be it; and the prospect of an egalitarian society is an unrealistic mirage. Alternatively, if sexual selection is not true, it should not be left to die in secret, but should be explicitly discredited lest sexual selection remain on the books as an obstacle to social justice.

In 1976 when Richard Dawkins wrote *The Selfish Gene*, he was primarily debunking group selection while emphasizing individual selection as the more important route to evolutionary success. He was criticizing a familiar narrative at the time that accounted for the evolution of traits by appeals to group or species benefits. I agree with Dawkins that individual selection is more significant to evolutionary success than group selection, and I respect his contributions concerning that point. I, too, seek evolutionary explanations that rely on individual selection and do not appeal to group selection. Instead, my challenge concerns the scientific accuracy of the philosophical world view that the phrase "selfish gene" has come to represent.

For example, *The Selfish Gene* publicizes a view of nature emphasizing competition: "We are survival machines–robot vehicles blindly programmed to preserve the selfish molecules known as genes," Dawkins writes. He continues with, "Our genes made us. We animals exist for their preservation and are nothing more than their throwaway survival machines. The world of the selfish gene is one of savage competition, ruthless exploitation, and deceit." In *River Out of Eden*,[13] Dawkins writes, "The universe we observe has precisely the properties we should expect if there is, at bottom, no design, no purpose, no evil and no good, nothing but blind pitiless indifference." And he states in the book, *Devil's Chaplain*,[14] "Blindness to suffering is an inherent consequence of natural selection. Nature is neither kind nor cruel but indifferent." These books develop a philosophy of universal selfishness, conflict, and lack of empathy as though revealed through evolutionary biology.

13. Richard Dawkins, *River Out of Eden: A Darwinian View of Life* (Basic Books, 1995).
14. Richard Dawkins, *A Devil's Chaplain: Reflections on Hope, Lies, Science, and Love* (Houghton Mifflin, 2003).

The issue before us is not whether this philosophy is appealing or repugnant. The issue is whether it is a true and accurate account of nature, of what the birds and bees around us are doing—whether their lives are really selfish and filled with uncaring sexual conflict.

This book is planned for release in 2009, the year of what promises to be a grand bi-centennial celebration of Darwin's birthday on February 12, 1809. Yet, among the anticipated horde of books praising the Darwinian "triumph," one will be able to discern critiques of some parts of Darwin or of later work extending Darwinism.

One critique, associated with Eva Jablonka and Marion Lamb,[15] Mary Jane West-Eberhard,[16] Massimo Pigliucci,[17] and Sean B. Carrol[18] challenges the genetic assumptions of neo-Darwinism, not Darwinism itself. This challenge does not extend to Darwin's writings, because Darwin wrote before genes were discovered, and Darwinism does not take a position on the mechanism of inheritance. This challenge does argue that the "new synthesis" of Ernst Mayr[19] and others in the 1940s to 1960s, which applied neo-Darwinism to systematics and paleontology, is inadequate in light of today's molecular and developmental genetics. This critique arises from evolutionary-developmental biology, or evo-devo for short, and seeks to update neo-Darwinism—indeed Eva Jablonka told me she has at times described herself as a Darwinist, but not a neo-Darwinist.

15. Eva Jablonka and Marion J. Lamb, *Evolution in Four Dimensions: Genetic, Epigenetic, Behavioral, and Symbolic Variation in the History of Life* (Cambridge: The MIT Press, 2005); E. Jablonka, "From Replicators to Heritably Varying Phenotypic Traits: The Extended Phenotype Revisited," *Biology and Philosophy* 19 (2004): 353–375; cf. also: Richard Dawkins, "Extended Phenotype–But Not Too Extended. A Reply to Laland, Turner and Jablonka," *Biology and Philosophy* 19 (2004): 377–396.

16. Mary Jane West-Eberhard, *Developmental Plasticity and Evolution* (New York: Oxford University Press, 2003); cf. also: Eva Jablonka, "Genes as Followers in Evolution–A Postsynthesis Synthesis? A Review of Mary Jane West-Eberhard, Developmental Plasticity and Evolution, 2003," *Biology and Philosophy* 21 (2006): 143–154.

17. Massimo Pigliucci, *Phenotypic Plasticity: Beyond Nature and Nurture* (Baltimore: The John Hopkins University Press, 2001); Massimoo Pigliucci, "Phenotypic Integration: Studying the Ecology and Evolution of Complex Phenotypes," *Ecology Letters* 6 (3): 265–272 (2003).

18. S. B. Carroll, "Evolution at Two Levels: On Genes and Form," *PLoS Biol* 3(7): e245 doi:10.1371/journal.pbio.0030245 (2005).

19. Ernst Mayr, *Animal Species and Evolution* (Cambridge: Harvard University Press, 1963).

Although the new synthesis of the 1940s needs some far-reaching revision to take account of contemporary genetics and developmental biology, it has made a long-lasting and likely permanent contribution in changing how the biological categories are understood. The original biological classification system of Linnaeus envisioned that a species could be identified with a "type" specimen that could be taken as definitional. One could go to a research museum like the British Museum in London or the American Museum in New York, open a drawer, and pull out a specimen of say, a robin, and it would instantiate the definition of "robin-hood." Any other specimen of a bird could be compared to the type specimen to determine if it was a robin too. If the specimen didn't exactly match, it was imperfect and not quite a "true" robin. But since the new synthesis, a species has been identified with a *sample* from a population. The museum drawer now contains a dozen or more robins that collectively define robin-hood. So, an unknown specimen now is compared to the collective sample to see if it falls within the range of variation naturally exhibited throughout the sample. If the specimen fits within the sample, it's a robin, plain and simple.

This denial of any "type" specimen carries over to within-species categories, too. There is no type specimen of a race, gender, or sexuality profile. Even though the rejection of what Ernst Mayr termed *typological thinking* has been foundational to evolutionary biology for nearly 75 years now, this insight has yet to be fully appreciated in medicine, especially pediatrics, urology, psychology, and psychiatry. These medical disciplines persist in attempting to "diagnose" persons into human categories that to an evolutionary biologist's eye are purely social constructs without any possibility of biological reality or precision. In biology, nature abhors a category. In biology, nature consists of rainbows within rainbows within rainbows . . .

In contrast, my sexual-selection critique does go back to what Darwin wrote, and how he conceived of social relations between males and females. Nonetheless, its replacement, social selection, is an evolutionary theory, too. Sexual selection and social selection both lie under the common umbrella of evolutionary biology, but offer very different accounts of what it takes to achieve evolutionary success within the gene pool of any species that possesses a social system, which is all of them, including in a sense, plants.

Finally, I have also pressed a third critique that pertains to evolutionary biology generally, and not Darwin specifically, and that has been also made by many biologists over the years. Evolutionary biology awaits an extended discussion about what an "individual" is.[20] Who counts as an individual seems obvious when thinking of ourselves, or our pets, livestock, or other organisms who become detached from their parents at birth. But individuality is not so clear cut in many other species. A grove of poplar trees consists of many trunks springing from one seed—what is the individual, a single tree trunk or the entire grove? Or consider a strawberry plant, or beach grass on a sand dune—these reproduce with runners, either above or below the ground. What does survival of the fittest mean when we can't say exactly who it refers to—the fittest tree trunk or the fittest grove of trees?

Without explicitly saying so anywhere, biologists define an individual as an entity containing one genome in one body. The problems turn up in species where one genome gives rise to multiple bodies, as in poplar trees, strawberries, and beach grass, or where multiple genomes reside within a single body, as in the endosymbiosis of algae with fungi in lichens, of algae with corals, and of algae with giant coral-reef clams. And then there are clusters of distinct individuals who are bunched together into a single functional unit, such as the Portuguese Man-O-War jellyfish, which consists of many separate polyps attached to one another that float around together like a spaceship in the ocean. In this case, survival of the fittest means survival of the fittest cluster, which then trickles down somehow to the prosperity of the individual polyps within the cluster.

Defining individuality has long been a problem for mycologists.[21] A hypha is a thin tube surrounded by a wall. Typically, hyphae are divided

20. Joan Roughgarden, *Evolution and Christian Faith: Reflections of an Evolutionary Biologist* (Washington: Island Press, 2006), 71–78.

21. Alan Rayner and Nigel Franks, "Evolutionary and Ecological Parallels Between Ants and Fungi," *TREE* 2 (1987): 127–133; Alan D.M. Rayner, "The Challenge of the Individualistic Mycelium," *Mycologia* 83 (1): 48–71 (1991); A. D.M. Rayner, *Degrees of Freedom Living in Dynamic Boundaries* (Imperial College Press, 1997).

into cells by cross-walls called septa. Septa usually have pores in them large enough for ribosomes, mitochondria, and sometimes even nuclei to flow among cells. Therefore, a hypha is like a conduit through with all the subcellular players can traffic back and forth. A mycelium is a branching network of hyphae, and these are found in soil and on, or in, all the substrates where fungi live. Fruiting bodies like mushrooms might grow up out of the mycelium. A mycelium may be too small to see or may be gargantuan. A 2,400-acre site in eastern Oregon had a contiguous growth of mycelium estimated to be 2,200 years old.

Many phenomena are already known, and more are being discovered, about how fungi interact with one another when mycelia contact. They may fuse and exchange nuclei or they may repulse each other. The cells between septa may contain multiple nuclei that are identical to one another or that may be genetically different. It's hard to find a vocabulary to describe the physiology within, and interaction between, fungal mycelia. I like to compare a fungal mycelium to a termite nest. Perhaps you may have seen a termite nest outdoors together with the system of corridors that the termites construct out of mud. If you open a crack in one of the corridors, you can see termites moving back and forth through their tunnels. A mycelium is similar, but it's the nuclei and other cellular organelles that traffic back and forth. In this case, the nuclei can be regarded as the "individuals," but these individual share space and function with other types of individuals, the mitochondria, and other organelles. In comparison to fungi, the plants and animals that have only a single nucleus per cell are special cases, like one house per lot in suburban living rather than a duplex, townhouse, commune, and other living arrangements preferred across diverse human cultures and locales.

Biologists working with social insects, the bees, wasps, ants, and termites, have also long puzzled about what an individual means. The common picture of a social insect nest consisting of a single queen plus workers has long been explained by kin-selection theory. I can recall teaching this theory as though gospel during the 1970s, and it figures so prominently in *The Selfish Gene*. Worker-ants are daughters of the queen-ant and share a high genetic relationship to her because of a special genetic system that bees, wasps and ants happen to have (called haplo-diploidy).

Because of an unusually high genetic relationship between daughter and mother in these species, worker-ants can send copies of their own genes into the next generation more effectively *via* the route of helping their mother to produce offspring rather than bothering to produce offspring by themselves.[22] This idea is fine, provided a nest really does consist of one queen together with her daughters as the workers.

However, social insect nests in many species form a huge distributed system with often unrelated multiple queens and cohorts of workers with differing parentage.[23] An extreme example is the "super-colony" of the red wood ant on a coastal plain in Hokkaido, which has 306 million workers and over a million queens living in 45,000 interconnected nests across a territory of 2.7 km^2.[24] These distributed entities in which the individual is poorly defined defy any ready application of kin-selection theory to explain the often cooperative social dynamics. Kin-selection is not incorrect, merely inadequate.

Although the forms of life just mentioned are well known to invertebrate zoologists and botanists, and are well described in their textbooks, evolutionary biology and its textbooks simply avoid such species in which the definition of an individual is problematic. The seeming clarity with which evolutionary biology seems to bespeak of individualism in books like *The Selfish Gene* is illusionary because only a subset of the species is being addressed by evolutionary studies, namely, the subset in which the definition of an individual is not problematic. Belief that evolutionary science somehow reveals the naturalness of individualism is circular, because it is based on species chosen in part because they do not challenge an individualistic perspective.

22. W. D. Hamilton, "The Genetical Theory of Social Behavior. I. II," *J Theor Biol* 7 (1964): 1–52.

23. Bert Hölldobler and Edward O. Wilson, "The Number of Queens: An Important Trait in Ant Evolution," *Naturwissenschaften* 64 (1977): 8–15; Laurent Keller, "The Assessment of Reproductive Success of Queens in Ants and Other Social Insects," *Oikos* 67(1): 177–180 (1993); David C. Queller et al., "Unrelated Helpers in a Social Insect," *Nature* 405 (2000):784–786.

24. S. Higashi and K. Yamauchi, "Influence of a Supercolonial Ant *Formica (Formica) yessensis* Forel on the Distribution of Other Ants in Ishikari Coast," *Jap J Ecol* 29 (1979): 257–264; S. Higashi, "Polygyny and Nuptial Flight of *Formica (Formica) Yessensis* Forel at Ishikari Coast, Hokkaido, Japan," *Insectes Sociaux, Paris* 30 (3): 287–297 (1983).

We can't continue indefinitely to neglect or downplay species in which the definition of an individual is problematic. After all, there are many thousands of species like poplars, beach grass, mushrooms, giant clams, corals, tunicates, bryozoans, Portuguese Men-O-War, and so forth whose evolution we'll have to understand someday. But more interestingly for the purposes of this book, social connections are just as important as material connections. Sure, beach grass stems share a *material* connection through their roots. But members of a pack of wolves or a pod of whales share a *social* connection that binds them into a unit, a team, every bit as real as the material connections that bind popular trees together.

The lack of clarity about what defines an individual in biology brings us to the fundamental problem with the selfish-gene metaphor—it overlooks the issues of decomposability and teamwork.

Take two robins, a male and female, who build a nest together. An interview in *The Guardian* on February 10, 2003, describes Richard Dawkins who "wanders over to the other side of the room and returns with a bird's nest that he picked up in Africa. 'It's clearly a biological object.' His eyes light up. 'It's clearly an adaptation. It's a lovely thing.' He says that birds do not need to be taught to make nests, they are genetically programmed to do so." Whether a bird's ability to make a nest is genetically endowed is not the point. The point is whether one or two birds cooperated to make it. Dawkins sees the evolution of nest-building ability as another success story for some selfish genes. In fact, the nest results from the *relationship* developed by the male and female during courtship. Both bring twigs to build the nest. The success of the genes in either bird is zero if the other doesn't do their job, and so the success of the nest is attributable to the relationship of trust established between both birds. Therefore, the genes for nest-building do represent an evolutionary success story, but not success because of selfishness. The success of the nest-building genes in any one robin rests on their ability to work with the genes in another robin.

Moreover, the success of the nest is not decomposable to a sum of the twigs brought by each, because half a nest is useless. No one has yet figured out a useful way to decompose team achievements into individual contributions. Does anyone really think a pitcher is the one who wins a baseball game, forgetting about the hitters and fielders? It's the team who wins or not. In the 30 years since the selfish-gene metaphor has gained traction as a popularization of neo-Darwinian thought, it has yet to emerge as a scientifically operational concept because of the decomposability problem.

Still, at this point a selfish-gene advocate typically retorts that a robin who cooperates with another robin in building a nest is helping itself, and so can be thought of as selfish after all. But that vacates the meaning of selfish. "Selfish gene" and "successful gene" are not the same thing. Think of it this way. Suppose a species has 25,000 genes in it and suppose 200 of them increase in frequency from one generation to the next while another 200 of them decrease. Why? If the increase is due to natural selection and not chance fluctuation, then do the 200 increase because they cooperate more effectively with each another than the other 200 who decline, or do the 200 increase because they interfere with, harm, or outcompete the other 200 for scarce limiting resources? No one knows. It's an open empirical question whether effective cooperation among genes versus effective competition between them underlies most evolutionary progress. But to stipulate that evolutionary success equals selfishness means we can't ask the question of which, cooperation or competition, is the more common route to evolutionary success.

The explanation for why the male and female robins cooperate to build a nest together cannot be subsumed under kin selection, reciprocal altruism, or group selection.[25] The male and female robins are usually not brother and sister or another close relative, so kin selection does not apply. The female and male robins are not exchanging altruistic acts that directly help each other as individuals, so reciprocal altruism does not apply. And a robin's nest does not bud off other nests, so group selection does not apply either—pairs don't reproduce other pairs, or live or die as a pair.[26]

---

25. Group selection is the differential survival and/or reproduction of groups and is contrasted with individual selection, which is the differential survival and/or reproduction of individuals.

26. A more extended discussion of the relation between kin selection and group selection appears toward the conclusion of Chapter 9.

Instead, the male and female robin are cooperating because they have a shared interest—together they form what might be called an "evolutionary household," a kind of evolutionary team, and they share a common bank account of evolutionary fitness consisting of the genes from both contained in the offspring they jointly raise.

"Cooperative teamwork" is a distinct principle from the older ideas of kin selection, reciprocal altruism, and group selection. In kin selection and reciprocal altruism, each individual acts solely from their own viewpoint—the worker-ant exploits the queen-ant to produce eggs that carry her genes rather than bother producing the eggs herself. Meanwhile, producing eggs is the queen-ant's evolutionary objective as well. The worker and queen share no common goal; they merely enjoy an independent coincidence of two individual goals. In cooperative teamwork, the evolutionary payoff is earned by the team as a unit through coordinated actions by its members in pursuing a team goal, and the payoff from team success is then distributed, perhaps unequally, among the team members with no individual earning anything unless their team succeeds. Cooperative teamwork is not an independent coincidence of individual interests, but the acceptance of a team goal and working together to attain that goal. Yet, effectiveness at cooperative nest-building evolves through ordinary individual selection, not group selection, because the route to evolutionary success for each *individual* robin is through cooperation with the other. On the other hand, cooperative teamwork is not group selection either, but individual selection with evolutionary success as an individual being attained through coordinated activity in pursuit of a team goal.

Now let's pass to the critique of sexual selection and to the development of its replacement.

Cooperation and Teamwork

ONE Sexual Selection Defined

To get clear about exactly what sexual selection is, we need three levels of definition. First is what might be called sexual selection in the narrow sense—this is the specific picture of how males and females relate to each other that Darwin originally enunciated. Next is sexual selection as a system of related scientific propositions—this is a collection of amendments and extensions to what Darwin originally wrote. Finally, there is sexual selection as an area of research, as an occupation within biology—this is the evolutionary study of social behavior which usually, but not necessarily, takes Darwin's picture of social behavior as its starting point.

I think that both sexual selection in the narrow sense and sexual selection as a system of interlocking scientific propositions are incorrect. In contrast, I think sexual selection as a research discipline is alive and healthy, having itself uncovered the very information that undercuts

Darwin's original picture of male/female social behavior. Sexual selection as a research discipline in principle need not be based on Darwin's picture, but based on whatever field studies and theoretical analysis eventually discover. From this standpoint, even my theory of social selection can be considered part of the overall research program of sexual selection. Still, I think it's confusing to portray social selection as part of sexual selection, even if by this one means the sexual-selection research area. I think it's more accurate to distinguish sexual selection from social selection as an alternative hypothesis to explain reproductive social behavior, a task that falls within the research area of evolutionary behavioral ecology.

So just what is Darwin's picture, sexual selection in the narrow sense? Toward the end of his career Darwin focussed on traits like the peacock's tail and the deer's antlers, called male ornaments, which seemed to have no importance for improving survival.[1] Instead, Darwin turned to the possible role such traits might have for mating. Darwin theorized that male ornaments evolved because females prefer males on the basis of the ornaments they exhibit, and males compete with each other for access to females. Darwin wrote that females had an innate "aesthetic" for beautiful features and by choosing mates with these features, the females bred the males to be beautiful. The rationale for females to mate with ornamented males is that their own sons will be likewise endowed, and thus given an advantage when their time comes to mate. He wrote, "Many female progenitors of the peacock must... have... by the continued preference of the most beautiful males, rendered the peacock the most splendid of living birds."

Some ornaments might be nonfunctional, like the peacock's tail was assumed to be, whereas others might be weapons, such as antlers that males were thought to use in combat with the victor winning access to females. This would lead to the evolution of well-armed males, because those without effective armament would leave few descendants. Although male-male combat might conceivably circumvent female choice

1. C. Darwin, *The Descent of Man and Selection in Relation to Sex* (London: John Murray, 1871).

by limiting their choice only to the victors, females were assumed to prefer the winners anyway because the winners were, *ipso facto*, the best of the males and by mating with them females tended to ensure that their own male offspring would be well-armed and effective at acquiring mates too. Thus, female preference for victorious males caused males to become "vigorous and well-armed... just as man can improve the breed of his game-cocks by the selection of those birds, which are victorious in the cock-pit." All in all, males evolve to be beautiful *and* well-armed, because of female mate choice.

In Darwin's writing, the peacock's tail and the deer's antlers were emblematic of male-female social relations generally. Darwin wrote, "Males of almost all animals have stronger passions than females," and "the female... with the rarest of exceptions is less eager than the male... she is coy." The phrases "almost all" and "with rarest of exceptions" show that Darwin was enunciating a theory not solely for the gender roles in peacocks and deer, but for all of nature, excepting a few rarities we need not worry about.

Darwin is the source for the phrases "passionate male" and "coy female," which are supposed to apply generally to the sex roles throughout nature. For example, take a walk in the park. Look at any bird, frog, fish, beetle, or butterfly, any animal at all, and according to Darwin's sexual-selection narrative, the male of the species is supposed to be passionate and the female coy. In making these claims, Darwin was enunciating a "norm" for sexual conduct throughout nature. By setting up a norm, any social behavior that differs from it can be branded as an "exception," or even as a "deviance," but if there are enough exceptions, maybe the "norm" is not really a norm, but just one of the many possibilities.

This characterization of sex-role norms is not some quaint anachronism. Restated in today's biological jargon, the narrative is considered proven scientific fact. The geneticist, Jerry Coyne,[2] confidently declares, "We now understand . . . Males, who can produce many offspring with

---

2. Jerry Coyne, "Charm Schools," review of *Evolution's Rainbow,* by J. Roughgarden, *Times Literary Supplement* 30 July 2004.

only minimal investment, spread their genes most effectively by mating promiscuously… Female reproductive output is far more constrained by the metabolic costs of producing eggs or offspring, and thus a female's interests are served more by mate quality than by mate quantity." By "quality" Coyne and other biologists mean "genetic quality." So, the passionate male has become the promiscuous male and the coy female the constrained female. Yet, the spirit of this present-day narrative remains identical to Darwin's narrative of nearly 130 years ago. In today's jargon, males with cheap sperm are constantly on the make for chances to mate, and females are constrained by their expensive eggs to try to ascertain which males have the best genes so as not to dilute the genetic quality of their heavy investment in eggs. On this account, the males can be ranked in terms of their objective genetic quality, females try to identify the male highest in the genetic hierarchy, and settle for what they can get.

The Darwinian sexual-selection narrative must surely seem familiar, because it has been thoroughly assimilated into popular culture. For example, the column in *Elle* magazine, "Ask E. Jean," reports that "Males fighting for females is the elastic in the jockstrap of evolution, therefore women are hardwired to 'size up' and appreciate male competition."[3]

The Darwinian narrative does not postulate that females each have their own private genetic preference, for example, where Sally the peahen likes the genes in Fred the peacock, whereas Betty the peahen prefers the genes in Bill the peacock. No. Fred and Bill must have a common genetic ranking with Fred better than Bill, for example. Then both Sally and Betty must prefer Fred, and one will get Fred while the other will have to settle for Bill. That's what Darwin's female aesthetic is all about. In peacocks, for example, if each female had a private aesthetic, then some would like red tails, some blue, and so forth, leading to a diversity of peacock tails, whereas, according to Darwin, a common preference for the same type of tail has led all the peacocks to exhibit nature's most beautiful bird tail.

In fact, examples are well known where each female seems to have her own private aesthetic. Male bowerbirds of New Guinea and Australia build decorated structures called bowers that resemble thatched huts at

3. Ask E. Jean, *Elle*, Feb. 2005.

which they display to females. Females choose the males in whose bowers they deposit eggs on the basis of his bower. Because of the conspicuous role for female choice in bowerbird courtship, they have long been used as exemplars of sexual selection. But there is no single best bower. The most complex bower is made by males of the Vogelkop Gardener bowerbird that lives on five remote mountains of Indonesian New Guinea. Bower styles differ among species, among populations of a species, and between individuals of a population. According to Jared Diamond, a naturalist who has long worked in New Guinea and who may be familiar to many because of his best-selling book, *Guns, Germs and Steel*, these differences in bower preference among females arise from intrinsic differences among birds rather than local differences in objects available for decorating bowers.[4] Whenever females have private preferences for different males, then the sexual-selection narrative is actually violated, even though female choice is taking place, because there isn't one best male that all females prefer. (Indeed, the differences in bowers may be culturally inherited rather than genetically inherited.[5]) This situation occurs even more starkly in mate choice related to the histocompatibility locus in which compatibility of parental genes is needed for effective immune response in the offspring.[6]

The phenomenon of a private aesthetic as contrasted with a common aesthetic (sometimes called noncongruent preferences versus congruent preferences) leads to the distinction of female choice for "good genes" versus "compatible genes." Using this distinction, sexual-selection theory is amended to say that female choice of mates might be intended to supply her eggs with sperm bearing either good genes (a common aesthetic)

4. J. Diamond, "Animal Art: Variation in Bower Decorating Styles Amongst Male Bowerbirds *Amblyornis inornatus,*" *Proceedings of the National Academy of Sciences USA*, 83: 3042–3046 (1986); Gail L. Patricelli, J. Albert C. Uy, and Gerald Borgia, "Multiple Male Traits Interact: Attractive Bower Decorations Facilitate Attractive Behavioural Displays in Satin Bowerbirds," *Proc R Soc Lond B* 270: 2389–2395 (2003); J.R. Madden, "Interpopulation Differences Exhibited by Spotted Bowerbirds *Chlamydera Maculata* Across a Suite of Male Traits and Female Preferences," *Ibis* 148 (3): 425–435 (2006).

5. J.R. Madden et al., "Local Traditions of Bower Decoration by Spotted Bowerbirds in a Single Population," *Anim Behav* 68: 759–765 (2004).

6. L. Bernatchez and C. Landry, MHC Studies in Nonmodel Vertebrates: What Have We Learned About Natural Selection in 15 Years? *Journal of Evolutionary Biology*, 16: 363–377 (2003).

or compatible genes (a private aesthetic), depending on the species—good genes in peacocks and compatible genes in bower birds. Indeed, some authors claim that the good-genes contribution to fitness is small compared to the compatible-genes effect.[7]

The addition of the compatible-genes basis for female choice is certainly an improvement over the original Darwinian sexual-selection narrative that focused solely on a common ranking of male quality. Still, we need to be up front on what is going on here. Research has actually falsified the original Darwinian hypothesis for at least some species, and investigators have, therefore, widened the original sexual-selection hypothesis to accommodate the species that contradicted the original hypothesis. In this way, sexual-selection theory grows bigger and bigger, adding amendments as needed to accommodate more and more species that depart from the original sexual-selection theory in one way or another, and sexual selection becomes what I call a "system" of interlocking propositions.

Some modifying of the original sexual-selection hypothesis is reasonable and inevitable, especially because Darwin was writing so many years ago. But continuously widening sexual-selection theory converts it into a system that becomes increasingly hard to test and impossible to falsify, and so sexual selection slowly morphs from a scientific theory into a doctrine or ideology. How much widening is too much? This is difficult to say. The more widening that is needed to accommodate the data, the more one gets suspicious that the whole approach is on the wrong track to begin with.

So, the question before us is to what extent the original Darwinian narrative of male/female relationships, sexual selection in the narrow sense, is true. The original narrative is certainly plausible. But how often is it true? Is it, in fact, ever true? In how many species is the male really the passionate sex and the female the coy sex? In how many species do females case out the genes of potential partners, and then choose based on genetic quality, either for good genes or compatible genes, or for genes at all?

---

7. B.D. Neff and T. Pitcher, "Genetic Quality and Sexual Selection: An Integrated Framework for Good Genes and Compatible Genes," *Molecular Ecology* 14(1): 19–38 (2005).

TWO The Case against Sexual Selection

My reasons for thinking that sexual selection is completely on the wrong track appear at some length in *Evolution's Rainbow* as well as articles since then.[1] Here is a distillation of the key points. There are three classes of objections, three strikes. First are the great many species that depart from the sexual-selection templates of male and female behavior in one way or another. Second are the increasing number of species that appeared perfectly to fit sexual-selection theory, but who nonetheless do not support it

1. Joan Roughgarden, Meeko Oishi, and Erol Akçay, "Reproductive Social Behavior: Cooperative Games to Replace Sexual Selection," *Science* 311: 965–969 (2006); Joan Roughgarden, "Challenging Darwin's Theory of Sexual Selection," *Daedalus* 136 (2): 1–14 (2007); Joan Roughgarden, "Social Selection vs. Sexual Selection: Comparison of Hypotheses," in *Controversies in Science and Technology. Volume II: From Climate to Chromosomes*, eds. Daniel Lee Kleinman et al. (New Rochelle: Mary Ann Liebert, 2008), 421–463.

when closely studied. The failure of its "poster-child" species to accord with sexual-selection theory raises the question of whether the theory is correct for any species whatsoever. Third are contradictions between population-genetic theory and the sexual-selection narrative that need to be circumvented for the theory to be internally consistent.

Before going further, I need to emphasize that my critique of sexual selection does *not* question the existence of female choice, but questions Darwin's view of *what* females are choosing *for*. Sexual selection is not synonymous with female choice. When Darwin introduced his theory of sexual selection, an ensuing controversy questioned whether females in nature were capable of the choice Darwin envisioned.[2] Darwin writes that female choice "implies powers of discrimination and taste on the part of the female which will at first appear extremely improbable; but by the facts to be adduced hereafter, I hope to be able to show that the females actually have these powers."[3] This ancient dispute underlies why many still seem to equate sexual selection with the mere existence of female choice. But the issue we face today is different. Today's issue is not whether females in nature have the capability for making sophisticated choices. But in Darwin's picture, the objective of female choice is to breed with beautiful and well-armed males. Today's sexual-selection critique questions whether female choice is actually for that *objective*, namely, to find males whose traits indicate high genetic quality or whether the objective is some other consideration more likely to improve her evolutionary fitness.

DEPARTURES FROM SEXUAL-SELECTION "NORMS"

Let's begin with the many species that depart somehow from the norms prescribed by sexual- selection theory.

2. Alfred Russel Wallace, "Darwin's 'The Descent of Man and Selection in Relation to Sex,'" *The Academy* (March 1871): 177; Alfred Russel Wallace, "The Colours of Animals," *Nature* (July 1890): 289; Alfred Russel Wallace, "Note on Sexual Selection," *Natural Science* (December 1892): 749–750; cf. http://www.wku.edu/~smithch.

3. Charles Darwin, *The Descent of Man; and Selection in Relation to Sex*, 2nd ed. (New York: Crowell, 1874), 219.

Sexual Monomorphism

Sexual selection is intended to explain a difference between males and females—males with ornaments and armaments, and females drab. Sexual dimorphism is presupposed. In many species, males and females are virtually indistinguishable, such as the guinea pigs many people raise as pets, or birds like penguins, where sexes can only be distinguished by careful inspection of the genitals. Sexual-selection theory is silent on why species vary from the highly sexually dimorphic peacock and to the sexually monomorphic penguin. Thus, sexual monomorphism is problematic for sexual selection.[4]

Mating Initiation and Frequency

In species where males do have a different appearance from females, males may not be passionate nor females coy. Female alpine accentors from the central Pyrénées of France, for example, solicit males for mating every 8.5 minutes during the breeding season. Approximately 93% of all solicitations are initiated by the female approaching the male, the other 7% by him approaching her.[5]

Sex-Role Reversal

In sea horses and pipe fish, the male is drab and the female ornamented, and male seahorses raise the young in a pouch into which the female deposits eggs, making males "pregnant," so to speak. Indeed, among fish species in which parental care is provided, it is usually the male who provides it, not the female,[6] and a male contribution to parental

---

4. N. E. Langmore and A. T. D. Bennett, "Strategic Concealment of Sexual Identity in an Estrildid Finch," *Proc R Soc Lond B* 266: 543–550 (1999).

5. N. B Davies et al., "Female Control of Copulations to Maximize Male Help: A Comparison of Polygynandrous Alpine Accentors, *Prunella collaris*, and Dunnocks, *P. modularis*," *Anim Behav* 51: 27–47 (1996).

6. J. D. Reynolds, N. B. Goodwin, and R. P. Freckleton, "Evolutionary Transitions in Parental Care and Live Bearing in Vertebrates," *Phil Trans R. Soc Lond B* 357: 269–281 (2002); cf. also Elizabet Fosgren et al., "Unusually Dynamic Sex Roles in a Fish," *Nature* 429: 551–554 (2004).

care is common in birds.[7] Some seahorses, pipefish, and birds, such as the jacana and phalarope, exhibit what biologists call "sex role reversal." The females are showy and males drab, the reverse of the peacock. Thus, sex role does not follow directly from gamete size with cheap sperm causing male passion and expensive eggs necessitating female coyness. Male seahorses make tiny sperm just as male peacocks do, and female seahorses make large eggs just as peahens do, but nonetheless male seahorses care for the young and female seahorses entrust their eggs to a male's pouch.[8]

Template Multiplicity

In many species, multiple types of males and females, each with distinct identifying characteristics, carry out special roles at the nest, both before and after the mating takes place. In the sandpiper-like European ruff, black-collared males defend small territories called courts within a communal display area called a lek. Meanwhile, white-collared males accompany females while the females feed. The white-collared males then leave the company of the females and fly to the lek where they are solicited by the black-collared males to join them in their courts. When the females eventually arrive at the lek to mate, they encounter pairs of males—one black-collared male paired with one white-collared male in some courts, as well as single black-collared males in courts by themselves. Evidently, females prefer to

7. D. Lack, *Ecological Adaptations for Breeding in Birds* (London: Chapman and Hall, 1968); R. Pierotti and Annett CA, "Hybridization and Male Parental Investment in Birds," *Condor* 95: 670–679 (1993); M. C. McKilrick, "Phylogenetic Analysis of Avian Parental Care," *Auk* 109: 828 (1993); Ellen D. Ketterson and Val Nolan Jr., "Male Parental Behavior in Birds," *Annu Rev EcoL Syst* 25: 601–628 (1994); A. Cockburn, "Prevalence of Different Modes of Parental Care in Birds," *Proc Roy Soc B* 273: 1375–1383 (2006).

8. T. Clutton-Brock, and A. C. J. Vincent, "Sexual Selection and the Potential Reproductive Rates of Males and Females," *Nature* 351: 58–60 (1991); Amanda Vincent et al., "Pipefishes and seahorses: are they all sex role reversed?" *TREE* 7: 237–241 (1992).

mate with a pair of black-collared and white-collared males rather than solely with one black-collared male.[9]

I have conjectured that the white-collared males have formed bonds with the females while accompanying them during their feeding. White-collared males may serve as "marriage brokers" who introduce females to the black-collared males who have not previously had the opportunity to meet females while they were busy setting up and defending courts in the leking area.

Regardless of whether my conjectures about this example are correct, cases such as these mean there is no single template for all the males in a species and another template for all the females. I have termed such distinct templates within a sex as *genders* and would say that there are two genders of males in the ruff. Actually, there are three male genders in ruffs, as we now note.

Transgender Presentations

In some species with multiple genders of males, one of the genders more or less resembles the females, as in bluegill sunfish in the lakes of North America.[10] The ruffs just mentioned possess a third male gender that has

---

9. J.G. van Rhijn, "Behavioural Dimorphism in Male Ruffs *Philomachus pugnax* (L.)," *Behaviour* 47: 153–229 (1973); J.G. van Rhijn, "On the Maintenance and Origin of Alternative Strategies in the Ruff *Philomachus Pugnax*," *Ibis* 125: 482–498 (1983); J.G. van Rhijn, "A Scenario for the Evolution of Social Organization in Ruffs *Philomachus pugnax* and Other Charadriiform Species," *Ardea* 73: 25–37 (1985); D.B.Lank et al., "Genetic Polymorphism for Alternative Mating Behaviour in Lekking Male Ruff *Philomachus Pugnax*," *Nature* 378: 59–62 (1995); D. Hugie and D. Lank, "The Resident's Dilemma: A Female Choice Model for the Evolution of Alternative Mating Strategies in Lekking Male ruffs (*Philomachus pugnax*)," *Behav Ecol* 8: 218–225 (1997); F. Widemo, "Alternative Reproductive Strategies in the Ruff *Philomachus Pugnax*: A Mixed ESS?" *Anim Behav* 56: 329–336 (1998).

10. M.R. Gross, "Sneakers, Satellites and Parentals: Polymorphic Mating Strategies in North American Sunfishes," *Z Tierpsychol* 60: 1–26 (1982); M.R. Gross, "Evolution of Alternative Reproductive Strategies: Frequency-dependent Sexual Selection in Male Bluegill Sunfish," *Phil Trans R. Soc Lond B* 332: 59–66 (1991); W.J. Dominey, "Female Mimicry in Bluegill Sunfish—A Genetic Polymorphism?" *Nature* 284: 546–548 (1980); W.J. Dominey, "Maintenance of Female Mimicry as a Reproductive Strategy in Bluegill Sunfish (*Lepomis macrochirus*)," *Env Biol Fish* 6: 59–64 (1981).

neither a black or white collar—it has no collared ornament at all, just like the females.[11]

In addition, male sunangel hummingbirds of the Andes from Venezuela to Bolivia have feathers called a gorget, a broad band of distinctive color on the throat and upper chest. Museum specimens reveal that some sunangel hummingbird species include "masculine females" with gorgets just like the male.[12] Also, some males have feminine coloration. All in all, of 42 studied species, seven have both masculine females and feminine males, nine have masculine females and no feminine males, two have feminine males and no masculine females, and 24 have neither masculine females nor feminine males.

Such natural variation in gender expression as well as gender multiplicity contradict the Darwinian prescription of one template, or norm, per sex—passionate, armed, and ornamented males on one side versus coy, drab females on the other. Such natural variation in gender expression points to multiple norms for each sex, a multiplicity of possible roles.

Homosexual Mating

In over 300 species of vertebrates, same-sex sexuality has been documented in the primary peer-reviewed scientific literature as a natural component of the social system.[13] Examples include species of reptiles like whiptail lizards, birds like the pukeko of New Zealand and the European oystercatcher, and mammals like giraffes, elephants, dolphins, whales, sheep, monkeys, and one of our closest relatives, the bonobo chimpanzee. A museum in Oslo has recently initiated exhibits that introduce

11. J. Jukema and T. Piersma, "Permanent Female Mimics in a Lekking Shorebird," *Biol Lett* 2: 161–164 (2006).

12. R. Bleiweiss, "Widespread Polychromatism in Female Sunangel Hummingbirds (*Heliangelus*: Trochilidae)," *Biol J Linnean Soc* 45: 291–314 (1992); R. Bleiweiss, "Asymmetrical Expression of Transsexual Phenotypes in Hummingbirds," *Proc R Soc Lond B* 268: 639–646 (2001).

13. B. Bagemihl, *Biological Exuberance: Animal Homosexuality and Natural Diversity* (New York: St. Martin's Press, 1999). Bagemihl, largely unknown to biologists, received his Ph.D. in linguistics from the University of British Columbia.

the public to the extent of homosexuality in nature. The museum earned international coverage and acclaim for its courage to present what the Discovery Channel and other nature-show media avoid in their laundered accounts of animal social life. Homosexuality in animal societies, as well as the large number of heterosexual matings often needed for a conception to take place, cast doubt that the sole function of sexuality in nature is to fertilize eggs with sperm. Instead, extensive homosexuality in nature is consistent with the hypothesis that the natural function of most sexuality is to sustain bonds between animals, bonds comprising the social system within which offspring are reared.

The cornucopia of "exceptions" to the original Darwinian templates of male-female behavior, that is, to sexual selection in the narrow sense, has necessitated a plethora of corollaries and amendments as workarounds that now form the sexual-selection system. Special stories have now been developed for why female alpine accentors from France are more passionate than males, why sex-role reversal occurs, why multiple genders (called "alternative mating strategies") occur, and why homosexuality occurs. The proponents of sexual selection think that these extensions are each and individually correct, or if not, and if one turns out to be wrong, then any particular extension can be fixed up without repercussions to the overall opus. The proponents of sexual selection may be right that all, or almost all, of the extensions and corollaries are correct, but maybe not.

Dime-Stories

A common theme runs throughout all the sexual-selection extensions and corollaries. This theme paints a picture of animals whose whole lives revolve around lying, cheating, and stealing from one another. To experience hell, one need not descend to the bowels of the earth, it is sufficient to be reincarnated as a nonhuman animal here on the earth's surface. The plight of a male in any arbitrary species is to fight with other males to control access to females, or if unable to win at combat, to make the "best of a bad job," as sexual-selection advocates say, by coercing a female to have sex with him against her wish and out of sight of the male who possesses her, or to masquerade as a female and surreptitiously mate with his harem, or to submit to homosexual mountings to tire the controlling male and then mate with his harem. And what about the female? Well, she may entrap a male to remain monogamous with threats to mate elsewhere or keep him busy with her passionate sexuality, or she may succumb to sexual violence and coercion directed against her because she "wins by losing," as one biologist put it. And so on. The diversity of special stories manufactured to account for real-life diversity in gender expression, sexuality, and family structure among animals reads like an orgy of dime-store novels.

Vocabulary Poison

And these stories employ a dime-store novel's vocabulary to trivialize and pathologize any departures from the sexual-selection norm. In the primary peer-reviewed literature, males are described as being "cuckolded," females as "faithful" or "promiscuous," offspring as "legitimate" or "illegitimate," males who do not hold territory as "floaters" or "sneakers" (code for "sneaky fuckers") all of whom are "sexual parasites," small males as "gigolos," feminine males as "female mimics" or even as "transvestite serpents" or "she-males" (a pornographic reference), and so forth. The cesspool of adjectives invented ostensibly as descriptions of animal behavior makes locker-room banter seem genteel. This vocabulary poisons any aspiration to scientific objectivity for today's sexual-selection system.

Sexual-selection advocates resist a critique that emphasizes what they regard as exceptional species, because such a critique ignores the supposedly many cases in which they claim the sexual-selection narrative is correct—a critique focused on "exceptions" throws the baby out with the bath water. Well, is there any baby in the bath water or just floating debris? Sexual-selection theory doesn't work even with the species for which it was designed, as is becoming increasingly clear from huge long-term studies. I now turn to several studies that have appeared since 2004 when *Evolution's Rainbow* was published. These studies pertain to poster-child species that were supposed to exemplify sexual selection rather than species that depart in some obvious way from sexual-selection expectations.

UNCONFIRMED "POSTER CHILD" SPECIES

Collared Flycatcher

The collared flycatcher is a small migratory woodland bird of central and eastern Europe studied for over 24 years on the Swedish island of Gotland, yielding a cumulative sample of over 8,500 individually marked individuals.[14] (Reflect on how much work this has involved.) This species would appear to be a perfect exemplar of classic sexual-selection theory. The males have a conspicuous white spot on their foreheads, which is called a "badge." The background to a landmark report in 2006 sets the stage for what was hoped would be a definitive demonstration of the sexual-selection narrative. The investigators write, "previous research has documented the importance of the male flycatcher's ornament (a conspicuous white forehead patch) in sexual selection through male-male competition, sperm competition, and mate choice. Sons inherit their fathers' forehead patch size. Females prefer to mate with males with a large forehead patch, especially late in the season, and receive fitness benefits from doing so."

But what did these investigators actually find? First, they confirm that male badge size does have a statistically significant heritability of moderate magnitude.[15] Fine so far, but now the show-stopper. The major problem is a near absence of heritability in male fitness.[16] That is, although males may vary among themselves in how many offspring they sire, almost none of this variation can be inherited, rendering it pointless for a female to attempt to ascertain which males have the best genes. Not surprisingly then, there is almost no heritability for female choice of

---

14. Anna Qvarnström, Jon E. Brommer, and Lars Gustafsson, "Testing the Genetics Underlying the Co-evolution of Mate Choice and Ornament in the Wild," *Nature* 441: 84-86 (2006).

15. h^2 (male-badge-size) = 0.38 ± 0.03.

16. h^2 (male-fitness) = 0.03 ± 0.01.

male badge size. That is, if a female does happen to prefer males with large badges, this preference is not inherited among her daughters.[17] Therefore, these negligible heritabilities result overall in a zero genetic correlation between female choice and male badge size,[18] the *sine qua non* of sexual selection. Hence, the investigators conclude that "genes coding for mate choice for an ornament probably evolve by their own pathways instead of 'hitchhiking' with genes coding for the ornament." Thus, preference by females for the badge does not endow their sons with genes that will cause them to be desired as mates in the future.

Blue Tit

Let us turn next to another study that appeared in 2006, this time with the blue tit, a woodland bird in the United Kingdom.[19] The blue caps on heads of males and females appear to be the same when viewed in regular visible light, but differ greatly when seen in ultraviolet light. As with the collared flycatcher, previous studies had set the stage for believing that the head cap served as an ornament in males to indicate genetic quality so that, as the investigators write, "females can evaluate the fitness of a male's offspring according to the reflectance properties of his crown."

If this hypothesis is true, then the authors "predict that ornaments should have a high heritability and that strong positive genetic covariance should exist between fitness and the ornament." To test this prediction the authors used 3 years of cross-fostering data from over 250 breeding attempts and partitioned the covariance between parental coloration and aspects of nestling fitness into a genetic and environmental component.

Contrary to the prediction of sexual-selection theory, the authors found that "variation in coloration is only weakly heritable"[20] and that

17. h^2(female-choice-of-badge-size) = 0.03 ± 0.01.

18. r (female-choice-to-male-badge) = −0.02 ± 0.17.

19. Jarrod D.Hadfield et al., "Direct Versus Indirect Sexual Selection: Genetic Basis of Colour, Size and Recruitment in a Wild Bird," *Proc R Soc B* 273: 1347–1353 (2006).

20. h^2(sire-cap-color) = 0.10 ± 0.11, which is not significantly different from zero.

"two components of offspring fitness—nestling size and fledgling recruitment are strongly dependent on parental effects, rather than genetic effects. Furthermore, there was no evidence of significant positive genetic covariation between parental colour and offspring traits." Specifically, they found that the genetic correlation between male cap color and whether the fledglings returned to breed the next year to be 0.40 ± 0.40, which is not significantly different from zero; the genetic correlation between male cap color and offspring head length to be 0.06 ± 0.11, which again is not significantly different from zero; and the genetic correlation between male cap color and offspring tarsus length to be −0.22 ± 0.10, which also isn't significantly different from zero.

Hence, the authors offer a remarkably direct summary of their conclusions: "Models of indirect sexual selection predict that sexually selected traits should have high heritability, that the magnitude of genetic variance in fitness should be substantial, and that there is significant positive genetic covariation between the sexually selected trait and fitness. In contrast to these predictions, this study has demonstrated that chromatic variation in the cap and chest of the blue tit is only weakly heritable, that variation in chick recruitment is determined to a large degree by environmental, rather than genetic effects, and that the genetic covariation between colour and fitness components is either non-significant or negative. Taken together, these results suggest that neither cap colour nor chest colour are likely to accurately reflect any genetic benefits a female may gain by mating to highly ornamented males."

Barn Swallow

Then in the next year, 2007, a study appeared on another bird species, the barn swallow, which had been extensively discussed as an exemplar of sexual selection.[21] You're probably familiar with the profile of a swallow, with their characteristic V-shaped tail and two trailing streamers. On

21. Jakob Bro-Jørgensen, Rufus A. Johnstone, and Matthew R. Evans, "Uninformative Exaggeration of Male Sexual Ornaments in Barn Swallows," *Current Biology* 17: 850–855 (2007).

average, the tails of males are longer than those of females. The tail length in females is taken to be aerodynamically optimal, and the elongated tail in males has been hypothesized to be exaggerated because of sexual selection—the longer tail is supposed to indicate an enhanced capability that comes from possessing good genes. In sexual-selection theory, the exaggerated tail length is called a "handicap," and its presence is hypothesized to indicate genetic quality—a handicap is supposed to reveal genes so good that a male can sustain the handicap and still function well. If so, females can choose males with the longer tails and thereby gain superior genes for their offspring.

The issue raised in this 2007 study is whether variation in tail length among males does indicate differing genetic quality. Here are the two possibilities: The sexual-selection scenario requires a single aerodynamically best tail length with the males differing in how much each is handicapped in accordance with their genetic quality. Or a multiplicity of tail lengths could be aerodynamically optimal corresponding to various body proportions, and if so, the difference among males would represent variation in what's aerodynamically best for each, plus a constant gender-identity label tacked on. In this case, differences in tail length are not handicaps that indicate differences in genetic quality.

As background, the authors acknowledge that, "over the last twenty years, a large amount of work has shown that female barn swallows are influenced by male tail length when choosing a mate." Then they state their finding: "We show that the aerodynamically optimal tail length varies significantly between males, whereas the extent of streamer elongation beyond the optimum does not . . . Therefore, contrary to handicap models of sexual selection, the sexually selected exaggeration of this trait provides females with little information about any aspect of mate quality." They write, "it is also worth reiterating that although there is exaggeration in the length of the tail streamer beyond the aerodynamic optimum, this does not vary significantly between individual males. This means that if female swallows use tail length as a cue to discriminate between males, they are likely to be doing so on the basis of variation in the underlying naturally selected trait and not on the extent of ornamentation beyond this optimum."

Lark Bunting

In early 2008, still more studies appeared that are problematic for sexual selection pertaining to species that appeared not to be exceptional. The lark bunting is a migratory songbird that breeds on prairies of the Great Plains of North America. Females are brown with dull white wing patches. Males are more colorful—generally black or black with patches of brown feathers, and have conspicuous white wing patches that vary in size and whiteness. Lark buntings were studied in the short-grass prairie of Colorado over 5 years from 1999 to 2003.[22] The authors found that the traits correlated with reproductive success varied across years, and that female choice did as well.

In successive years, the males that sired the most young, including both the young in their own nest plus those they sired in adjacent nests, were respectively: those with a bigger beak size, more black feathers on the body, whiter wing-patch color, smaller body size, and more black feathers on the rump.[23] Thus, the trait that marks the best male varies across years.

Meanwhile, in successive years, the male trait marking those with whom the female raised the most young were respectively: bigger beak size, bigger wing-patch size, more black feathers on the rump, bigger beak size, and fewer black feathers on body.[24] Thus, the trait that marks the male with whom the female can raise the most young varies across years.

Except for the first year, 1999, the best male from a male perspective—the one who sires the most young, isn't the best male from a female perspective—the one with whom she can fledge the most young. So, if females choose males for their genes to endow their sons with those genes, then female choice should align with the males who sired the most young. Alternatively, if females choose males to maximize the number of young they rear, their choice will align with the males with whom they

22. Alexis S. Chaine and Bruce E. Lyon, "Adaptive Plasticity in Female Mate Choice Dampens Sexual Selection on Male Ornaments in the Lark Bunting," *Science* 319: 459–462 (2008).
23. Op. cit., Fig. 2
24. Op. cit., Fig. S2 in supplemental online material.

produce the highest number of fledglings. In other words, is female choice for "quality" or quantity?

The authors assume that females determine which males get to breed because males are present in excess. Whether a male had a chance to mate turned out to correlate in successive years, respectively, with: bigger beak size, bigger wing-patch size, whiter wing-patch color, smaller wing-patch color, and blackness of feather color,[25] again showing variation across the years. Overall the results are not strong. However, taking all the traits together, the authors argue that female choice tends to align more with traits that predict maximizing the number of eggs she will fledge rather than with traits that predict whether the male is a successful sire. This result would agree with experimental studies from Sweden with fish called the sand gobies where it was demonstrated that female gobies preferred males who were "good fathers" rather than males who were competitively superior over other males.[26]

The usual sexual-selection narrative clearly cannot apply in the lark bunting. As the authors state, "theory on the evolution of ornamental male traits by sexual selection assumes consistency in selection over time." However, they show that "in lark buntings sexual selection on male traits varied dramatically across years and, in some cases, exhibited reversals in the direction of selection for a single trait." This finding is difficult to accommodate into a sexual-selection narrative.

The lark bunting is not alone. Two other studies appeared later in 2008 describing fluctuating directions of selection: a wild population of Soay sheep on the island archipelago of St. Kilda, NW Scotland, in the North Atlantic[27] and a population of red jungle fowl at a field station of the University of Stockholm.[28]

25. Op. cit., Fig. 4.

26. E. Forsgren, "Female Sand Gobies Prefer Good Fathers Over Dominant Males," *Proc R Soc Lond B* 264: 1283–1286 (1997).

27. Matthew R.Robinson et al., "Environmental Heterogeneity Generates Fluctuating Selection on a Secondary Sexual Trait," *Current Biology* 18: 751–757 (2008).

28. Charlie K. Cornwallis and Tim R. Birkhead, "Plasticity in Reproductive Phenotypes Reveals Status-specific Correlations Between Behavioral, Morphological, and Physiological Sexual Traits," *Evolution* 62: 1149–1161 (2008).

Peacocks and Peahens

Also in early 2008, a remarkable study appeared refuting the standard sexual-selection narrative for the ultimate poster child of sexual selection, the peacock and peahen.[29] The authors state that the elaborate train of the peacock "is thought to have evolved in response to female mate choice and may be an indicator of good genes." Nonetheless, after studying a feral population of Indian peafowl in Japan for over 7 years, the authors conclude that: "We found no evidence that peahens expressed any preference for peacocks with more elaborate trains . . . similar to other studies of galliforms showing that females disregard male plumage. Combined with previous results, our findings indicate that the peacock's train (1) is not the universal target of female choice, (2) shows small variance among males across populations and (3) based on current physiological knowledge, does not appear to reliably reflect the male condition."

In addition to presenting their own data, they review the many studies that report contradictory results with peacocks, with those in the United Kingdom generally supporting the sexual-selection narrative, while those elsewhere do not. They comment that "positive results are likely to be published and distributed in the research field of sexual selection" and caution that "it is equally important to publish negative results," thus raising suspicion of a publication bias favoring sexual selection in behavioral-ecology research.

The authors also offer a novel suggestion about how the evolution of peacock tails might be framed. They write, "peacock-like plumage is inhibited by a high level of female hormones, suggesting that there has been selection on females for dull-coloured plumage, as would be expected in ground-nesting species with little or no male parental care where females suffer high predation risk during incubation" and note that "peahens were actually twice as vulnerable as males in our population (24 female versus 11 male adult birds were predated during our

---

29. Mariko Takahashi et al., "Peahens do not Prefer Peacocks with More Elaborate Trains," *Animal Behaviour* 75: 1209–1219 (2008).

study, even though the sex ratio of the population was generally male biased"). They then develop an alternative perspective: "Phylogenetic studies have suggested that oestrogen dependent plumage dichromatism such as that in Indian peafowl was preceded by bright monochromatism in both sexes, followed by natural selection for duller coloration in females... in contrast with the conventional explanation of sexual selection in which current plumage sex differences are predominantly the result of selection for greater ornamentation in males, rather than selection for lesser ornamentation in females." I find it stunning that researchers on peacocks and peahens are no longer considering the male's tail to be a derived specialization to attract females, but instead are regarding the female's lack of a tail to be a derived specialization to avoid predators, such as wolves and foxes.

Indeed, dimorphism resulting from the female evolving cryptic coloration to avoid predation, rather than the male evolving bright coloration to advertise his quality to females, is actually common and vastly underpublicized. Later in 2008, a study appeared summarizing the evolutionary routes to dimorphism in swallowtail butterflies.[30] As long ago as 1865, Wallace observed that in many swallowtail butterflies of the Malay peninsula, mimicry occurred only in females, not males.[31]

Mimicry is when one species closely resembles another. During their caterpillar stage, while feeding on leaves, butterflies can accumulate the toxic chemicals plants use to defend against herbivores. Some butterfly species thereby become inedible themselves. Thereafter, some perfectly edible species may evolve to resemble the inedible ones. This mimicry whereby edible species copy the appearance of inedible ones is called Batesian mimicry, after the 19th century naturalist Henry Bates.

Wallace proposed that females, and not males, evolve Batesian mimicry because females, due to their very heavy egg loads and less effective escape flights, face greater predation risk and gain a greater advantage

30. Krushnamegh Kunte, "Mimetic Butterflies Support Wallace's Model of Sexual Dimorphism," *Proc R. Soc B* 275: 1617–1624 (2008).

31. A. R. Wallace, "On the Phenomena of Variation and Geographical Distribution as Illustrated by the Papilionidae of the Malayan Region," *Trans Linn Soc* 25: 1–71 (1865).

through mimicry compared with males. Thus, according to Darwin, the male and female primitively share the same drab appearance, and then the male evolves a brighter color. According to Wallace, the male and female primitively share the same bright appearance and then the female evolves an appearance to protect her from predators, which is either to look drab and cryptic or to mimic an inedible species. The newly published study using molecular phylogenetic methods confirmed Wallace's version—in butterflies that are sexually dimorphic, the female coloration is usually derived, whereas the male coloration retains the ancestral pattern.

Incidentally, coloration in butterflies reveals still more complications. In many edible species, the females are polymorphic with various morphs resembling different inedible species and even a morph retaining the ancestral appearance shared with the males. These females, which look like males, have been called "transvestites,"[32] as though the female appearance were primitive and the male appearance a later derivation. In fact, the male-appearing females might simply represent a fraction that has not evolved the Batesian mimicry, which would be expected if the abundance or toxicity of the inedible species that is being copied were unreliable in space or time.

The study on swallowtail butterflies was predated in 2006 by a similar investigation of dimorphism in 240 species of the dragon lizards that are distributed across Africa, Asia, and the Australian-Papuan region.[33] Ornaments in these lizards include throat sacs, dorsal crests, fleshy spines, and other structures on the head. The authors found a trend whereby ornaments usually evolve in both sexes at the same time. Then in some lineages, ornaments are lost in females, resulting in sexual dimorphism wherein only the male remains ornamented. The authors report that the loss of ornaments is associated with predator avoidance. They write that

32. C. Clarke et al., "Male-like Females, Mimicry and Transvestism in Butterflies (Lepidoptera: Papilionidae)," *Syst Entomol* 10: 257–283 (1985).

33. T. J. Ord and D. Stuart-Fox, "Ornament Evolution in Dragon Lizards: Multiple Gains and Widespread Losses Reveal a Complex History of Evolutionary Change," *J Evol Biol* 19: 797–808 (2006).

"lizards relying on soil burrows and/or burying themselves in sand either lose or are less likely to evolve protruding ornaments," that "species occupying more open habitats, where predation risk is expected to be higher, are significantly less likely to evolve (or retain) conspicuous colouration or ornamentation than lizards found in closed environments," and that "habitat openness explains more interspecific variance in female ornamentation than males, suggesting that the influence of natural selection is potentially more pronounced for females."

In all these studies, the trend is for dimorphism to result from the loss in females of ornaments initially present in both sexes. Why then were the ornaments present to begin with? They are evidently used for communication between and within the sexes. The loss of ornaments by females might, therefore, indicate a reduction in male-female and female-female communication to avoid attracting predators (run silent, run deep), and any remaining male displays would have nothing to do with advertising genetic quality, but would represent the limited communication that can be safely carried out in dangerous circumstances.

Extra-Pair Parentage

Still more direct evidence against sexual selection has come from studies of what are called extra-pair matings. A "pair" in this context refers to a male and female who jointly tend eggs in a nest—the set of eggs in the nest is called a "clutch." A clutch usually contains a few eggs whose parents differ from the birds tending the nest, but instead is a male or female from a nearby nest. This "extra-pair parentage" results from "extra-pair copulations." Sexual-selection theorists postulate that extra-pair copulations represent "cheating" by a female who is paired with a genetically inferior male and who seeks a genetic upgrade by mating with an adjacent male whom she ascertains to have better genes than her nest-partner, called her "pair-male." For example, the authors of the collared-flycatcher study just discussed report that "forehead patch size has also been directly implicated in a good-genes process because extra-pair offspring that are sired by highly ornamented males fledge in relatively better condition and therefore have higher survival chances than within-pair offspring." We have already seen data showing that female choice of the pair-male in collared flycatchers cannot be explained as a quest for badge-size genes with which to endow her sons. But could the good-genes rationale nonetheless still apply to the choice of extra-pair males? Biologists hope so. Research on extra-pair parentage might reveal the importance of indirect genetic benefits without the "complication" of any direct ecological benefits that the pair-male may provide because all the extra-pair male is presumably providing a female is his genes. It's worth noting though that the possibility that extra-pair males do in fact provide some form of direct benefit, such as protection or risk sharing, is never checked. By stipulation, the extra-pair male is said to provide only his genes.

Erol Akçay, who received his Ph.D. from my laboratory at Stanford in June 2008 and whose research I'll be presenting in Chapters 8 to 11, has compiled a survey and meta-analysis of over 100 studies of extra-pair paternity in birds spanning over 50 species.[34] He finds that only 40% of the studies report some kind of indirect genetic benefit (either good genes or compatible genes) accruing to females who pursue extra-pair matings, and the other 60% actually found that genetic benefits of any type were absent. These statistics might suggest that females pursue extra-pair paternities to obtain genetic benefits in a minority of cases, but not in general. Yet, even this qualified conclusion is probably too generous. The null hypothesis here is that genes are irrelevant to female choice of mates, either within-pair or extra-pair. Therefore, by coincidence, 50% of the time whatever females are choosing for in their males happens to coincide with a genetically determined marker like a badge, and 50% of the time it doesn't coincide with a genetically determined marker. Furthermore, a statistical meta-analysis shows that extra-pair males are on average not different than within-pair males in traits that are supposed to indicate the genetic quality of males, such as plumage color or singing behavior. Finally, the meta-analysis also showed that extra-pair offspring don't survive better than within-pair offspring, showing that whatever the females are looking for, it does not confer any survival benefits to their offspring. Their choice might, of course, increase the number of young they are able to raise, but not the survival ability of those young. So, the statistics on

34. Erol Akçay and Joan Roughgarden, "Extra-pair Paternity in Birds: Review of the Genetic Benefits," *Evolutionary Ecology Research* 9: 855–868 (2007).

extra-pair paternity studies can be understood as indicating that genetic benefits are completely irrelevant to female choice.

The collared flycatcher, blue tit, barn swallow, lark bunting, and peacock are poster-child species for sexual selection. And yet, detailed studies of these species refute a sexual-selection explanation for their male ornaments—the badge in the collared flycatcher, the UV head cap in the blue tit, the tail in the barn swallow, the wing patch in the lark bunting, and train in the peacock. Similarly, a great many studies of extra-pair parentage in birds also fail to offer strong evidence, or any evidence at all, for a sexual-selection interpretation. Nonetheless, the literature of evolutionary biology contains thousands of papers that claim to support sexual selection. If one were simply counting paper titles, sexual selection would appear to be a well-established principle, and therefore, species departing from the sexual-selection narrative can legitimately be treated as exceptions. But evidently, sexual selection can never be demonstrated once studies are sufficiently thorough. This development raises the question of whether there are any species anywhere that truly manifest the sexual-selection narrative.

The realization that sexual selection might be entirely incorrect is undoubtedly disturbing to sexual-selection advocates, and an initial reaction to these recent studies might be to return to the early literature in which sexual selection was supposed to have been demonstrated long ago. Of the classic studies on sexual selection, none has been more influential than Bateman's 1948 laboratory experiments with *Drosophila*.

Bateman Experiments

In 1948, the English geneticist, Angus Bateman, presented a study that has long been regarded as experimentally confirming Darwin's theory of sexual selection.[35] Bateman reported that, for a male, "fertility is seldom likely to be limited by sperm production but rather by the number of inseminations or the number of females available to him." Similarly, he claimed to have found in his flies an "undiscriminating eagerness in

35. A. J. Bateman, "Intrasexual Selection in Drosophila," *Heredity* 2: 349–368 (1948).

males and discriminating passivity in females" in accord with the sexual-selection narrative. More specifically, Bateman reported that male fitness (number of eggs bearing the male's paternity) increased with the number of mates, whereas for females, fitness was independent of the number of mates beyond one—one male was sufficient to supply all the sperm needed to fertilize the eggs. Furthermore, the males were observed to have a higher variance in fitness than the females, that is, some males had paternity of many eggs and some of very few, whereas the females all produced about the same number of eggs and so had about the same fitness as one another. This observation was interpreted to say that the strength of sexual selection is stronger in males than in females, and later authors even went on to say that males are, therefore, more highly evolved than females. Bateman's claims are referred to collectively as "Bateman's Principle" or the "Darwin-Bateman Paradigm" by various authors today. The Bateman experiments are a cornerstone of sexual-selection theory, and they have been widely cited in papers and textbooks in recent decades.

But in 2005, critiques of both Bateman's work itself, as well as how accurately it has been subsequently cited, appeared.[36] The critique by Zuleyma Tang-Martinez and Ryder[37] notes that Bateman presented data in two graphs. Only one "shows female fertility peaking after one mating." The other graph "shows that both male and female fertility increases as a result of siring young with more than one mate, although the slope for males is greater." In stating his conclusions, Bateman emphasized the results from only one of his two figures. Without ever monitoring behavior, he concluded that "these results were due to ardent, indiscriminate males and passive, choosy females." But Bateman couldn't observe how many matings actually took place; all he could observe were genetic

---

36. Critiques by Tang-Martinez, Dewsbury, and Gowaty and their colleagues are discussed here. See also J. D. Reynolds, "Animal Breeding Systems," *Trends in Ecology and Evolution* 11: 68–72 (2006) and P. A. Gowaty and S. P. Hubbell, "Chance, Time Allocation, and the Evolution of Adaptively Flexible Sex Role Behavior," *Integr Comp Biol* 45: 931–944 (2005).

37. Zuleyma Tang-Martinez and T. Brandt Ryder, "The Problems with Paradigms: Bateman's Worldview as a Case Study," *Integr Comp Biol* 45: 821–830 (2005).

markers of the males in the offspring, not the matings themselves. As Tang-Martinez and Ryder noted: "In reality, Bateman had no way of knowing how many times a female mated, or with how many males, because he did not conduct behavioral observations."

Tang-Martinez and Ryder refer to the subsequent "dogmatization" of Bateman's findings. They observe that key researchers in the field (Trivers,[38] Wilson,[39] Daly and Wilson[40]) as well as textbook authors (Alcock,[41] Drickamer et al.,[42], Krebs and Davies[43]) present data only from the graph that is claimed to support sexual selection, while ignoring the other. "Thus, only one portion of Bateman's experiments has been used to formulate one of his best known 'principles': that male RS [reproductive success] increases with number of mates, while female RS peak's after mating with only one male. Moreover, in some cases, Bateman's data and methodology have been misrepresented and embellished when doing so strengthened preconceived notions of male and female behavior."

Tang-Martinez and Ryder note further that "Bateman's use of *D. melanogaster* also is controversial because females of this species can store and use sperm for up to 4 days (Bateman's experiments lasted 3 or 4 days). Since in other *Drosophila* species, females need to copulate more frequently, Birkhead[44] concludes: 'Had Bateman chosen a species that typically recopulates more than every 3 or 4 days, Trivers would not have been able to disregard those results which did not fit his preconceptions about sexually motivated males and reluctant females.'"

38. R.L. Trivers, "Parental Investment and Sexual Selection," in *Sexual Selection and the Descent of Man*, ed. B. Campbell (Chicago: Aldine Publishing, 1972), 136–179.

39. E.O. Wilson, *Sociobiology: The New Synthesis*, (Cambridge: Harvard University Press, 1975).

40. M. Daly and M. Wilson, *Sex, Evolution, and Behavior* (Boston: Willard Grant Press, 1983).

41. J. Alcock, *Animal Behavior: An Evolutionary Approach* (Sunderland, Sinauer Associates, 1989).

42. L.C. Drickamer, S.H. Vessey, and E.M. Jakob, *Animal Behavior: Mechanisms, Ecology, Evolution* (New York, Mc Graw Hill, 2002).

43. J.R. Krebs and N.B. Davies, *An Introduction to Behavioural Ecology* (Oxford, Blackwell Scientific Publications, 1993).

44. T.R. Birkhead, *Promiscuity: An Evolutionary History of Sperm Competition* (Cambridge: Harvard Univ. Press, 2000), 198.

Tang-Martinez and Ryder continue, "Among Bateman's most important popularizers is Trivers,[45] whose influential paper on parental investment perpetuated the stereotypes of indiscriminate males and sexually restrained females. Trivers assumes that male and female parental and sexual behaviors can be explained by anisogamy [difference in size between the egg and sperm] and that females always start with a greater investment because eggs are costly but sperm are cheap. At any point in time, females will have invested more than males; consequently, females will safeguard their past investment by continuing to invest in parental care, while males will benefit by deserting their partner to seek additional mates. Thus, females will be predisposed to take care of the young and remain monogamous, while selection will favor promiscuous males." But, "Contrary to Dawkins'[46] statement that 'the word excess has no meaning for a male,' the antiquated notion that males can produce virtually unlimited numbers of sperm at little cost is demonstrably incorrect. Sperm depletion, as a result of previous ejaculations, has been reported in diverse invertebrate and vertebrate taxa."

Tang-Martinez and Ryder conclude by saying that they "suggest that traditional views emphasizing male promiscuity, female sexual passivity, and greatly differing costs of male and female reproduction do not provide the best framework for understanding mating system evolution and sexual selection."

A critique by Dewsbury makes the same points:[47] "Bateman apparently made only casual observations of the actual behavior of his flies (pp. 356–357). I can find no basis in Bateman's[48] article for Thornhill and Alcock's[49] statement 'all males exhibited courtship behavior and every

45. R. L. Trivers, "Parental Investment and Sexual Selection," in *Sexual Selection and the Descent of Man*, ed. B. Campbell (Chicago: Aldine Publishing, 1972), 136–179.

46. R. Dawkins, *The Selfish Gene* (Oxford: Oxford Univ Press, 1976), 176.

47. Donald Dewsbury, "The Darwin-Bateman Paradigm in Historical Context," *Integr Comp Biol* 45: 831–837 (2005).

48. A. J. Bateman, "Intra-sexual Selection in Drosophila," *J Hered* 2: 349–368 (1948).

49. R. Thornhill and J. Alcock, *The Evolution of Insect Mating Systems* (Cambridge: Harvard University Press, 1983).

female was vigorously courted' (p. 55) or for Trivers'[50] statement that the females that 'failed to copulate' (which should read 'failed to produce progeny') 'were apparently courted as vigorously as those who did copulate' (p. 137) or that the 21 percent of males who failed to reproduce showed no disinterest in trying to copulate, only an inability to be accepted (p. 138). I find no such observations reported."

Dewsbury concludes that the "Darwin-Bateman paradigm has been of great heuristic value. It probably holds in more cases than it fails. However, given its origins and the significant exceptions to its hypotheses, reevaluation is important. This is likely to yield a perspective that is both more complex and more comprehensive. I think it likely that when these proposals are seen in their cultural context and when further research is carefully conducted, we may get a very different understanding of males and females."

Then in 2007, still another critique of Bateman's work appeared by Snyder and Gowaty[51] who bent over backwards to avoid direct criticism. Near the beginning, they wrote: "We do not intend this reanalysis as a criticism of Bateman; his approach was groundbreaking and we are well aware that he accomplished his work without modern computational aids." While noting many statistical and data-analysis flaws, they also observe some arithmetic mistakes in one of Bateman's tables. They say, "We easily attributed two of the mistakes to rounding errors, but the other appears to be an error in arithmetic—acceptable in an era before calculators." The abacus was invented long before the calculator. Still, the authors state of Bateman that "his methods had flaws, including the elimination of genetic variance, sampling biases, miscalculations of fitness variances, statistical pseudo-replication, and selective presentation of data. We conclude that Bateman's results are unreliable, his conclusions are questionable, and his observed variances are similar to those expected under random mating." Nonetheless, the authors qualify their stunning conclusions by reiterating that regardless of their analysis, they

50. R. L. Trivers, "Parental Investment and Sexual Selection," in *Sexual Selection and the Descent of Man 1871–1971*, ed. B. Campbell (Chicago: Aldine, 1972), 136–179.

51. Brian Snyder and Patricia Adair Gowaty, "A Reappraisal of Bateman's Classic Study of Intrasexual Selection," *Evolution* 61: 2457–2468 (2007).

"do not intend this article as a criticism of Bateman; he accomplished his work without modern computational tools." The authors end with a "call for repetitions of Bateman's study using modern statistical and molecular methods" and feel that Bateman's work should be regarded in the future merely as "hypotheses" rather than established fact.

I admire the thoroughness of these reviewers and appreciate their circumspection. Still, I wonder whether the delicacy of phrase apparent in their conclusions serves the public's need to know the truth. These three critiques of Bateman and later workers have been devastating. There's simply no justification for continued adherence to the sexual-selection narrative on the basis of Bateman's work. Imagine if one spoke so guardedly of the belief, once widely held, that the earth is flat. Imagine a scientific review article in some royal court at the time of Christopher Columbus saying, "The hypothesis that the world is flat has been of great heuristic value and has stimulated the voyages of our brave explorers, but it may not provide the best framework for Principal Supplicants (PS's) on prospective research petitions. Therefore, we should carefully conduct many more rigorously designed blue-water oceanographic cruises in research galleons employing state-of-the-art canvas-sail technology with the latest analog magnetic-compass instrumentation. We should also convene cross-cutting interdisciplinary science advisory panels to facilitate the redirection of future research funding into peer-reviewed crusades to uncover the historical context of flat-earth science. Then we will be able to synthesize the flat-earth and round-earth theories into a consensus understanding of the earth's geometrical status within celestial-systems science." If these three independent critiques are not enough to reject Bateman's principle now and forever more, what would it take? Wouldn't it be more honest to set the beer mug down on the bar and just say it, "The earth ain't flat and Bateman's principle ain't so."?

POPULATION-GENETIC CONTRADICTIONS

Why do detailed studies of mate choice in birds fail to confirm sexual-selection theory? In my opinion, it's because sexual-selection theory is

48   COOPERATION AND TEAMWORK

contradictory to population genetics. One problem is simply that if females continually select males with the best genes, then bad genes are
quickly eliminated, so after several generations, when all males have become genetically equivalent, females no longer need to bother selecting
males on the basis of their genes.

Paradox of the Lek

This difficulty has been attracting increasing attention in the sexual-selection
literature where it has become known as the "paradox of the lek." A 2007
article[52] describes it thus: "Over time directional selection should erode the
genetic variation for secondary-sexual traits [like the peacock's tail], so that
females will no longer profit from discriminating among males based on
these traits and such female preferences should eventually disappear. Yet,
females continually display strong preferences for males with relatively
elaborate traits. This situation has been called the 'lek paradox.'"

This 2007 paper situates itself among nine other theoretical proposals
since 1990 offering different rationales for "resolving" the paradox.[53]

52. C. W. Miller and A. J. Moore, "A Potential Resolution to the Lek Paradox Through
Indirect Genetic Effects," *Proc R Soc B* 274: 1279–1286 (2007).

53. The following summary of citations is taken from Table 1 of Miller and Moore, op. cit.:
No cost to mate choice coupled with hidden nongenetic benefits: J. D. Reynolds and M. R.
Gross, "Costs and Benefits of Female Mate Choice: Is There a Lek Paradox?" *Am Nat* 136:
230–243 (1990); higher mutational input and selection for modifiers as a result of prolonged
directional selection: A. Pomiankowski and A. P. Møller, "A Resolution of the Lek Paradox,"
Proc R Soc B 269: 21–29 (1995); genic capture through condition-dependent expression of traits:
L. Rowe and D. Houle, "The Lek Paradox and the Capture of Genetic Variance by Condition
Dependent Traits," *Proc R Soc B* 263: 1415–1421 (1996); lower variance in male-mating success
than expected: R. V. Lanctot et al., "Lekking Without a Paradox in the Buff-breasted Sandpiper,"
Am Nat 149: 1051–1070 (1997); mechanisms of sexual selection result in apparent balancing se-
lection: A. J. Moore and P. J. Moore, "Balancing Sexual Selection Through Opposing Mate
Choice and Male Competition," *Proc R Soc B* 266: 711–716 (1999); genotype-by-environment
interactions (context-dependent mate choice): P. David et al., "Condition-dependent Signalling of Genetic Variation in Stalk-eyed Flies," *Nature* 406: 186–188 (2000) and F. Y. Jia, M. D.
Greenfield, and R. D. Collins, "Genetic Variance of Sexually Selected Traits in Waxmoths:
Maintenance by Genotype x Environment Interaction," *Evolution* 54: 953–967 (2000); mistake-
prone mate choice: J. P. Randerson, F. M. Jiggins, and L. D. Hurst, "Male Killing Can Select for
Male Mate Choice: A Novel Solution to the Paradox of the Lek," *Proc R Soc B* 267: 867–874
(2000); multivariate genetic variation orthogonal to direction of sexual selection: E. Hine, S. F.
Chenoweth, M. W. Blows, "Multivariate Quantitative Genetics and the Lek Paradox: Genetic
Variance in Male Sexually Selected Traits of *Drosophila serrata*," *Evolution* 58: 2754-2762 (2004).

Some authors propose elaborate genetic models postulating that secondary sexual characters are polygenic and incur very high mutation rates thereby accumulating a steady supply of bad genes to discriminate against. Such propositions have not been tested, much less validated.

The 2007 study just cited offers the most ingenious proposal to date. The authors hypothesize that beautiful ornaments in a male do not indicate anything about the male himself, but indicate instead that the male has a good mother, because the health of the offspring depends on maternal ability. A male presumably carries genes for his mother's maternal capability; therefore, a female should choose a beautiful male as a mate to endow her daughters with good maternal capability. Although this clever narrative gladdens my heart, it doesn't solve the paradox of the lek. It replaces that paradox with another one which might be termed the "paradox of the creche." If females are selecting beautiful males because beauty indicates genetically determined maternal capability, then after several generations all the females should be equally good mothers, again erasing any grounds for females choosing their mates on the basis of genes.

A 2008 article in this genre acknowledges that "the paradoxical persistence of heritable variation for fitness-related traits is an evolutionary conundrum that remains a preeminent problem in evolutionary biology."[54] The authors conclude "heritable variation is not eroded by continuous directional selection because, rather than leading to fixation of favored alleles, selection leads instead to allele frequency cycling due to the concerted coevolution of the social environment with the effects of alleles." No evidence whatsoever exists for a social-environment driven process of gene-frequency cycling.

Another 2008 article acknowledges that "directional female mate choice is expected to deplete additive genetic variation in male traits. This should preclude such trait-based choice from resulting in genetic benefits to offspring, and yet genetic benefits are the explanation for the choice. This evolutionary conundrum is known as the lek paradox."[55] The "resolution"

54. W. Edwin Harris, Alan J. McKane, and Jason B. Wolf, "The Maintenance of Heritable Variation Through Social Competition," *Evolution* 62: 337–347 (2008).

55. Janne S. Kotiaho et al., "On the Resolution of the Lek Paradox," *Trends in Ecology and Evolution* 23: 1–3 (2008).

to the paradox offered by this article amounts to a total capitulation: "The only resolution of the lek paradox is that the directional selection imposed by female choice is not enough to deplete the genetic variance in fitness." Sure. If female choice isn't effective at selecting for good genes, then it won't deplete the genetic variance in fitness. No problem.

The growing literature of the last dozen years aimed at "resolving" the paradox of the lek is itself testimony that the paradox has not been resolved. Perhaps it can't be resolved. Perhaps the paradox of the lek is a fatal flaw to sexual-selection theory.

Indeed, the collared flycatchers exhibit precisely what one would expect if the paradox of the lek is genuine. The existence of a moderate heritability for male badge indicates that it is not being subject to much directional selection because if it were, its heritability would have dropped to near zero because natural selection would have exhausted the additive genetic variance needed for evolutionary progress. However, the heritability of *fitness* is indeed near zero among flycatcher males, indicating that male flycatchers have already been selected to be well adapted genetically to their environment, and there is no additive genetic variation left for fitness. Therefore, the male badge is meaningless as an indicator of good genes, because all the genes are equally good by now, and so no inherited female preference for the badge is sustained through evolution. The collared flycatcher data set is exactly what one would expect if sexual selection is absent, in accordance with what has been anticipated by the paradox of the lek.

The lek paradox, although increasingly well-known, actually understates the magnitude of the population-genetic difficulties with sexual selection. The lek paradox allows that bad genes in males may have existed to begin with, but presumes they are eliminated through female choice that weeds out the bad genes. Because this paradox is "merely" a logical self-contradiction, hope springs eternal that it will someday be evaded through clever theorizing. A more fundamental difficulty apparently not recognized by sexual-selection proponents challenges the very plausibility of whether bad genes ever existed to begin with that are detectable through female choice.

Limited Scope for Fitness-Based Choice

Consider a recent study by John Byers, a program director of the behavioral ecology panel at the National Science Foundation, who oversees government funding of research projects pertaining to sexual selection. His study purports to demonstrate sexual-selection by females for male good genes among pronghorn antelopes in North America.[56] This species happens to be the second-fastest land mammal second only to cheetahs and capable of running up to 60 mph. Both females and males have horns of variable size with female horns usually being smaller.

Female pronghorns place a lot of time and energy into selecting the males who sire their young. The investigators write that "in the 2 weeks preceding estrus, each female visits several potential mates that hold widely spaced harems. Males cannot coerce or force copulation and do not offer resources. Each female moves independently and switches between harems . . . leaving males that fail to show effective harem defense. . . . This female sampling behavior is energetically expensive." The investigators report that the characteristics favored by females are "running speed, endurance, agility, and tactical spatial sense." As a result of this choosing, females "converge on a small proportion of males that sire most young."

The benefit of selecting these particular males is that the "offspring of attractive males were more likely to survive to weaning and to age classes as late as 5 years, resulting in a selection differential, calculated by expected differences in lifetime number of offspring weaned, of 0.32 against random mating." They explain: "We assigned fitness as expected lifetime number of offspring surviving to weaning and calculated fitness for a female that mated randomly and for a female that actively sampled and mated with a preferred male... Fitness for random mating was 3.53 and fitness for sampling and active choice was 5.15, yielding a selection differential of 0.32 against random mating. The size of the selection differential explains why energetically expensive mate sampling is maintained in pronghorn."

56. John A. Byers and Lisette Waits, "Good Genes Sexual Selection in Nature," *Proc Nat Acad Aci USA*, 103: 16343–16345 (2006).

The investigators conclude: "Preferred males, those that have demonstrated impressive vigor under the relentless scrutiny of the female mate sampling process, likely are those with relatively low numbers of small-effect deleterious mutations." Therefore, "female choice may be important and unrecognized as a force that can lower population genetic load."

Female choice is being represented as a kind of natural eugenics—its action cleanses the gene pool of small-effect deleterious mutations that have accumulated, reducing the "genetic load" that these mutations place upon the species.

The investigators are confident about the general implications of their claims. They begin their paper by stating, "The reality of active female mate sampling and mate choice is now unquestioned," and conclude their abstract by saying "female choice may be important and unrecognized as a force that can lower population genetic load."

This theme that the genetically best males are naturally entitled to sire the most offspring resulting in species betterment is a recurring feature of the sexual-selection narrative.[57] The issue before us is not whether it is appealing or repugnant that female choice might supply a universal natural system of eugenics, the issue is whether this claim is true. Is it true that female choice cleanses the gene pool of small-effect deleterious mutations?

No doubt the gene pool does accumulate small-effect deleterious mutations, but can female choice cleanse it? This present-day focus on the gene pool's bad genes, in this instance on weakly deleterious mutations, represents the school of thought in genetics that can be traced to the founder of eugenics, Francis Galton.[58] During the 1950s, studies claimed that every person has three to five recessive lethal genes that would cause their children to die if they chose the wrong marriage partner.[59]

57. Cf. Jacek Radwan, "Effectiveness of Sexual Selection in Removing Mutations Induced With Ionizing Radiation," *Ecology Letters* 7: 1149–1154 (2004).

58. Francis Galton, "Hereditary Talent and Character," *Macmillan's Magazine* 12: 157–166, 318–327 (1865).

59. H. Muller, "Our Load of Mutations," *Amer J Hum Genet* 2: 111–176 (1950); N. Morton, J. Crow, and H. Muller, "An Estimate of the Mutational Damage in Man from Data on Consanguineous Marriages," *Proc Nat Acad Sci USA* 42: 855–863 (1956).

This school has always been pessimistic about the evolutionary future, feeling that evolution has reached its pinnacle and any variation is useless or harmful.[60] The focus on deleterious mutations in the 1950s was consistent with the U.S. Atomic Energy Commission's funding of studies about radiation from nuclear bombs and power plants. This school is now sounding the alarm about the increase of deleterious genes in humans resulting from the relaxed selection associated with medical advance.[61]

But, even though small-effect deleterious genes exist in the gene pool, can female choice of males do anything about it? Does the sample of males a female encounters offer any scope for choice based on male fitness? Is there enough difference between the fitness of a genetically good male and a genetically bad male for discernment of fitness difference to be possible? If not, then female choice cannot impact on the gene pool's quantity of deleterious mutations.

Small-effect deleterious genes reside in the gene pool in a mutation-selection balance—they are being replenished through mutation, and eliminated by natural selection, leading to a steady state distribution. The distribution of the number of deleterious genes carried by individuals in a population is well-known,[62] and improved derivations are now available that use fewer assumptions and include possible gene-gene interactions.[63] The basic distribution is a roughly bell-shaped curve called a Poisson distribution. The location of the peak of this curve indicates the average number of deleterious genes in an individual. The peak's location has been derived as $L\mu/s$ where L is the number of genetic loci in the genome, μ is the probability that a deleterious mutation arises in an

60. See: quote of H. Muller, in M. Kimura and T. Ohta, *Theoretical Aspects of Population Genetics* (Princeton: Princeton University Press, 1971), 166; cited in R. Lewontin, *The Genetic Basis of Evolutionary Change* (New York: Columbia University Press, 1974), 30.

61. M. Lynch et al., "Perspective: Spontaneous Deleterious Mutation," *Evolution* 53: 645–663, 1999.

62. John Haigh, "The Accumulation of Deleterious Genes in a Population—Muller's Ratchet," *Theor Pop Biol* 14: 251–267 (1978).

63. Michael Desai, Daniel Weissman, and Marcus W. Feldman, "Evolution can Favor Antagonistic Epistasis," *Genetics* 177: 1–10 (2007).

individual at any particular locus, and s (called the selection coefficient) is the amount by which the fitness of an individual carrying a deleterious gene is reduced relative to an individual without a deleterious gene. That is, if we take 1 as the fitness of an individual without a deleterious gene, then $1 - s$ is how the fitness of an individual with the deleterious gene is denoted. If an individual has two deleterious genes, then its fitness is $(1 - s)(1 - s)$, indicating a double dose of deleteriousness. Similarly, if an individual as k deleterious genes, it's fitness is $(1 - s)^k$. So, with this formula, we can compute the average number of weakly deleterious genes in an individual using typical values for L, μ, and s. Let L be 25,000 loci, μ be 10^{-6}, and s be 0.001. Then the average number of weakly deleterious genes in an individual works out to be 25—typically, a different 25 genes for each individual, but still, 25 on the average. Now, what can female choice do about this?

Well, if the genetically average male has 25 weakly deleterious genes, how few does a genetically good male have, and how many does a genetically poor male have? To see the spread in male genetic quality, we consult the variance of the distribution which, being a Poisson, conveniently has the property that the mean equals the variance. So the variance in our example is also 25. The standard deviation then is 5 (i.e., $5 = \sqrt{25}$). So, a genetically good male has 20 deleterious genes, and a genetically bad male has 30 (i.e., the mean $\pm$ one standard deviation). Next, what is the fitness difference between this genetically-good male and genetically-bad male? Well, $(1 - s)^{20}$ for a genetically-good male works out to be 0.98 and $(1 - s)^{30}$ for a genetically-bad male is 0.97. These are nearly the same, a 1% difference. There's simply no way a female in the field can perceptually discern a one percent difference in the fitness of two males. Moreover, the sample size of males that a female will actually encounter is small, limiting her exposure to whatever genetic variety does exist. The huge fitness differences reported for pronghorn males between those who sire many and those who sire few cannot have anything to do with genes, and must somehow instead represent the outcome of social dynamics.

Still other self-contradictions lurk throughout sexual-selection theory. A recent review highlighted logical inconsistencies in the theory that supposedly connects parental investment to gamete size.[64] As the authors write, according to sexual-selection theory, "females are more committed than males to providing care because they stand to lose a greater initial investment. This, however, commits the 'Concorde Fallacy' as optimal decisions should depend on future pay-offs not past costs." The Concorde Fallacy refers to throwing good money after bad, as in the expensive European supersonic passenger plane, the Concorde, where increasing investments kept being justified because of the money already spent, rather than on the money to be earned in the future. The authors devise a work-around for this fallacy, but admit that even with their work-around, the argument "remains weak."

Another inconsistency the authors raise pertains to what follows from the premise that cheap-sperm males are present in excess relative to expensive-egg females. According to sexual-selection theory, this situation is supposed to imply that males should invest more time in competing with one another for access and control of females rather than investing in offspring care, or, as the authors put it, "I compete, I don't have the time to care." As the authors state, sexual-selection theory "assumes that the best response for males, who face more mating competitors than females, is to invest more heavily in weaponry, ornaments or other traits that increase their access to mates. There is, however, a valid counterargument: when the going gets tough, the smart do something else." Indeed, they cite empirical studies in which male parental care was interpreted as the way to avoid male-male competition, and other studies in which the absence of male parental care was interpreted as what was required to win at male-male competition. The authors state, "something is amiss if biologists can use the same logic to reach two diametrically opposed conclusions." The authors then develop some densely argued models in attempting to work around this dilemma.

In summary, considering the many diverse expressions of gender and sexuality that depart from the Darwinian sexual-selection "norms," the

64. Hanna Kokko and Michael D. Jennions, "Parental Investment, Sexual Selection and Sex Ratios," *Journal of Evolutionary Biology* 21: 919–948 (2008).

empirical studies that do not confirm the sexual-selection expectations even in the poster-child cases that seemed initially to exemplify sexual selection, and the presence of irreconcilable contradictions between population-genetic theory and the sexual-selection narrative, I have concluded that the entire theoretical system of sexual selection is incorrect and should be abandoned. As far as I'm concerned, because of these three strikes, sexual-selection is out. The trajectory of research since *Evolution's Rainbow* appeared has been to further undermine sexual selection theory, not to reinforce it, despite the often vehement condemnation that my opposition to sexual selection has evoked.

My case against sexual selection has been building since 2004 when *Evolution's Rainbow* appeared, and at this point readers may wonder what the other side has to say. An informed, constructive rebuttal by Tim Clutton-Brock to our *Science* paper[65] recently appeared, also in *Science*.[66] Clutton-Brock is a professor at University of Cambridge in the United Kingdom who has long worked on sexual selection and is perhaps the principal authority, or dean, of the subject. Therefore, it is particularly significant that his rebuttal concedes many of the points reviewed above, although he still concludes that sexual selection remains a valuable approach, a point to which I return. Here are some of the key concessions.

Concerning the supposed importance of gamete size in determining sex roles, Clutton-Brock acknowledges that "Sex differences in parental care are not an inevitable consequence of sex differences in gamete size because patterns of parental care are likely to co-evolve and feedbacks may be complex".

Concerning the supposed greater strength of sexual selection in males compared with females, Clutton-Brock writes "a substantial proportion of reproductive variance in both sexes is often caused by age, by random processes that do not contribute to selection, or by phenotypic differences that have no heritable basis," and adds that "relationships between

65. Joan Roughgarden, Meeko Oishi, and Erol Akçay, "Reproductive Social Behavior: Cooperative Games to Replace Sexual Selection," *Science* 311: 965–969 (2006).

66. Tim Clutton-Brock, "Sexual Selection in Males and Females," *Science* 318: 1882–1885 (2007).

relative reproductive variance in the two sexes and the evolution of sex differences are complex and inconsistent." The background to this statement is the idea that variance among members of the same sex in how much reproduction takes place, that is, whether some reproduce a lot while others reproduce slightly or not at all, measures the strength of sexual selection. It's often been asserted, going back to now-discredited 1948 studies of Bateman, that males have a higher variance in reproduction than females, implying that sexual selection is stronger in males than females. Clutton-Brock acknowledges that this long-held premise is dubious.

Concerning the supposition that males are more competitive than females because males must compete to acquire mates, Clutton-Brock writes the following: "Intense reproductive competition among females is not confined to species where males invest more heavily than females in their offspring." In other words, female-female competition occurs in nonsex-role-reversed species, too. He continues, "in a number of birds where females and males have similar ornaments, both sexes are commonly involved in aggressive displays with rivals." Specifically, "females more commonly compete with each other for access to resources necessary for successful reproduction (including breeding sites, parental care, and social rank) than for access to gametes produced by the opposite sex." Thus, Clutton-Brock acknowledges that even in nonsex-role-reversed species, females may compete as strongly with one another as males do with each other. According to Clutton-Brock, the only difference between male-male and female-female competition in nonsex-role-reversed species is that females compete for parental-care resources, whereas males compete for opportunities to mate. Thus, Clutton-Brock acknowledges that the central narrative of sexual-selection theory is often not accurate even for non-sex-role-reversed species, and meanwhile it remains inaccurate in sex-role-reversed species.

To accommodate females into an enlarged sexual-selection theory, Clutton-Brock develops a mirror-image narrative for female-female competition and mate choice by males for the best females. He writes that, "intrasexual competition between females for resources may generate large individual differences in fecundity that strengthen selection on males to identify and prefer superior partners and selection on females to signal temporal and individual differences in fecundity. Strong selection on females to maximize the growth and survival of their offspring may also generate selection pressures for mating with genetically compatible partners which, in some cases, may favor mating with multiple males." Yet, Clutton-Brock also admits that female-female competition is often mysterious. He writes, "where females compete directly with each other, it is often unclear precisely what they are competing for. Where females have developed obvious secondary sexual characters, it is often uncertain whether these are used principally to attract males or in intrasexual competition for resources, and how their development is limited is unknown. And, where males show consistent mating preferences for particular categories of females, we do not yet know whether they are usually selecting for heritable differences in female quality or for nonheritable variation in fecundity or for both."

The attempt to expand sexual-selection by developing a mirror image narrative for female-female competition and male choice of the best female, taking place simultaneously with male-male competition and female choice of the best male, is destined to fail for the same reasons already raised. Take the objections to the standard male-oriented narrative, do a global search and replace, and transpose every occurrence of "male" and "female," thereby producing a set of objections appropriate to a female-oriented mirror-image narrative. For example, if males are choosing the females genetically best at raising young, then soon the genetic variation for female quality will disappear, yielding the paradox of the creche, the mirror image of the paradox of the lek. Similarly, the genetic variation in female quality arising from accumulated weakly deleterious mutations will be less than 1%, implying a limited scope for male choice of female quality, just as there is limited scope for female choice of male genetic quality arising from weakly deleterious mutations. Trying to patch sexual-selection theory by giving females parity with males with regard to competition and mate choice is futile, and postpones coming to grips with the real prospect that sexual selection theory is fundamentally incorrect.

Clutton-Brock sums up by observing that "sexual selection is now commonly defined as a process operating through intrasexual competition for mates or mating opportunities, with the result that selection pressures arising from intrasexual competition between females to conceive or rear young are generally excluded and sexual selection is, by definition, a process that is largely confined to males." To incorporate females into sexual-selection theory, in addition to the males that are in the theory already, Clutton-Brock restates what sexual-selection is: "It may be helpful to return to a broader definition of sexual selection as a process operating through intrasexual competition for reproductive opportunities, providing a conceptual framework that is capable of incorporating the processes leading to the evolution of secondary sexual characters in both sexes."

Clutton-Brock's revised definition of sexual selection remains distinct from my alternative of social selection, as detailed further in the next chapter. The differences are first, social selection does not privilege "competition" as the principal intra-sexual dynamic because cooperation through teamwork is as important as competition. Second, social selection is not limited to intrasexual relationships, because teamwork exists between the sexes, such as bi-parental care, as well as within the sexes. Third, social selection understands that "reproductive opportunity" explicitly refers to the number of young successfully reared, and not to mating, which is the usual focus of sexual selection.

To conclude, Clutton-Brock asserts that "the theory of sexual selection still provides a robust framework that explains much of the variation in the development of secondary sexual characters in males, although the mechanisms controlling the relative intensity of reproductive competition and the relative development of secondary sexual characters in the two sexes are more complex than was originally supposed. The recognition of these complexities helps to refine the assumptions and predictions of the theory of sexual selection but does not undermine its basic structure."

Why? Why is sexual selection "still" a "robust framework"? Clutton-Brock acknowledges deep problems with the theory and then redefines the theory to sidestep the difficulties. Moreover, Clutton-Brock doesn't mention the many difficulties that lie beyond the immediate central narrative of sexual selection. Clutton-Brock doesn't address why sex-role reversal occurs, why multiple forms of males and females exist, why the presupposition of a clear-cut male-female binary is problematic, why homosexuality is not uncommon, what should be done about population-genetic contradictions such as the paradox of the lek and the limited scope for female choice of male genetic quality. Nor does Clutton-Brock cite the many failed poster-child species of sexual selection and other empirical studies that fail to confirm the sexual-selection narrative. And isn't the evidence that Clutton-Brock does cite sufficient by itself to call into question whether any more effort should be devoted to shoring up sexual selection theory? How much money, surely many millions of research dollars over the last two decades in the United States, United Kingdom, and Europe, has been invested in trying to confirm sexual selection, only to find that it needs to be redefined yet again? How many thousands of hours by scientists, volunteers, and students have been consumed in field and laboratory studies to confirm sexual selection only to be reassured that it is "still robust"? Isn't it time to cut losses and move on?

I hope that readers from social sciences such as psychology, anthropology, sociology, and international relations, and from humanities such as philosophy, ethics, and theology, who are interested in applying evolutionary biology to their disciplines will especially note Clutton-Brock's concessions. The sexual-selection narrative has been widely and uncritically adopted as axiomatic by these disciplines as a scientific and true account of biological nature. Workers in these disciplines must awaken to the realization that the sexual-selection area of evolutionary biology is not settled science, is in considerable flux, and is not ready for export. I hope, too, that journalists and science writers will withhold, or at least greatly qualify, their speculations and sound bites about the evolution of human social behavior.

This concludes my case against sexual selection—the prosecution rests.

THREE Social Selection Defined

The void left by the loss of sexual-selection theory presents us with a rare opportunity to carry out possibly long-lasting basic research on foundational issues in evolution—to rethink the science of sex, gender, and sexuality. Since *Evolution's Rainbow* appeared, my laboratory has been developing an alternative to the theoretical system of sexual selection that we term, *social selection*. A comparison of the central differences between social selection and sexual selection appears in Table 1.

To avoid repeating past mistakes, it's worth considering how sexual selection started off on the wrong foot. I suggest that Darwin initially erred by focussing on *quantity of mating*, which is only one component of evolutionary success rather than on the *quantity of offspring successfully reared*, which is the bottom line for evolutionary success. Obviously, offspring are impossible unless some mating takes place, but the quantity of

Table 1   Central Narratives of Social Selection and Sexual Selection

Social Selection	Sexual Selection
BEHAVIOR AS OFFSPRING-PRODUCING SYSTEM	BEHAVIOR AS MATING SYSTEM
Natural selection from differences in offspring-producing success. Males and females negotiate bargains and side-payments to control the social infrastructure and to maximize offspring production.	Natural selection from differences in mating success. Males compete for mating opportunities, females are a "limiting resource" for males, and females choose males for genes.

matings *per se* is only distantly related to quantity of offspring reared. Nonetheless, sexual-selection theory always refers to reproductive social behavior as comprising a "mating system." Within a mating system, evolutionary change then arises from differences in "mating success," and particular behaviors are understood by how they contribute to controlling and maximizing the frequency of mating. Male/female social dynamics are then seen to revolve around females as a "limiting resource" for males. Hence, males must compete with each other for access to mating opportunities with females, or for control of the females themselves, and females choose males to maximize the genetic quality of their offspring. Sexual selection errs by elevating a component of reproduction, namely mating, into an end in itself.

So, to take a different tack, social selection views reproductive social behavior as comprising an "offspring-producing system." Within an offspring-producing system, evolutionary change arises from differences in "offspring-producing success," and particular behaviors are understood by how they contribute to building, maintaining, and/or controlling the social infrastructure from which offspring are produced. Male/female social dynamics determine bargains and exchange side-payments

Table 2   Scope of Evolutionary Systems of Sex, Gender, and Sexuality

Evolutionary System of Sex, Gender, and Sexuality
Origin of Sexual Reproduction
Origin of Male-Female Binary
Modeling Social Systems—Number of Tiers
Central Narrative of Male/Female Social Dynamics
Subsidiary Narratives—Secondary Sex Characters, Conflict
Contradictions to Sexual-Selection—Sex Role Reversal
Peripheral Narratives for Sex/Gender Diversity
Application to Humans

between members of a mating pair, and between neighboring pairs, to manage the offspring-producing social infrastructure with the objective of maximizing the quantity of offspring placed into the next generation. I term these points contained in Table 1 as the *central narratives* of social selection and central selection.

As we've already seen, however, many additional issues pertain to sexual-selection theory. Table 2 illustrates the extent of the issues. Theories pertaining to each of these issues are associated with both social selection and sexual selection to produce what I term *evolutionary systems of sex, gender, and sexuality*. Each system contains not only its own central narrative, but also different hypotheses at each step in the full sweep of issues faced by both systems.

For example, both systems include hypotheses for the evolution of sex. Sexual selection favors hypotheses that stress a hypothetical role for sexual reproduction in pruning the gene pool of deleterious mutations or that envision an eternal evolutionary arms race between predator and prey and among competing species. Social selection emphasizes the role of sexual reproduction in continually rebalancing a species' portfolio of genetic variation to meet continually changing circumstances. Next, sexual selection favors hypotheses that view the origin of the male-female difference as resulting from a primordial sexual conflict. Social selection favors hypotheses that view the origin of male and female as a shared strategy to maximize the number of gametic encounters, thus maximizing the number of eggs that are successfully fertilized. And so forth, one by one, the social selection and sexual systems offer different accounts for the origin and maintenance of all the features of the genetics and behavior associated with sexual reproduction.

The theory of social selection proposed here did not spring forth fully formed. It draws upon many threads already in evolutionary biology, but also includes new ideas and weaves them all into a new system. It is also a work in progress, one to which all are invited. In later passages, I will discuss the preexisting threads and will do so from a personal perspective. Also, I'll introduce the members of our "social-selection project" as I discuss everyone's research. We'll get to these specifics in the next two parts. For now, and to conclude this chapter, I'd like to acknowledge prior uses of the phrase "social selection" and to distinguish my use from that of others.

The phrase, "social selection," has been previously used for a different concept by Mary Jane West-Eberhard.[1] She writes, "The special characteristics of sexual selection discussed by Darwin apply as well for social competition for resources other than mates." She describes social competition as "competition in which an individual must win in interactions or comparisons with conspecific rivals in order to gain access to some resource. The contested resources might include food, hibernation space, nesting material, mates, or places to spend the night. Seen in this broader perspective, *sexual selection* refers to the subset of social competition in which the resource at stake is mates. And *social selection* is differential success . . . in social competition." Thus, social selection *sensu* West-Eberhard is a generalization of the central competitive narrative of Darwin's sexual selection.

Allen Moore and coworkers[2] have since broadened West-Eberhard's definition by writing, "Selection reflecting associations between fitness

1. Mary Jane West-Eberhard, "Sexual Selection, Social Competition, and Speciation," *Quart Rev Biol* 58 (1983): 155–183.
2. Allen Moore et al., "The Evolution of Interacting Phenotypes: Genetics and Evolution of Social Dominance," *American Naturalist* 160 (suppl): S186–S197 (2002).

and social behavior has been termed 'social selection.'"[3] Continuing, "social selection occurs whenever a covariance exists between interacting phenotypes[4] such as covariances established as a result of nonrandom social interactions, behavioral modification in response to social interactions, indirect genetic effects, or relatedness among interactants." Similarly, Steven Frank, in a review of mathematical population-genetic theory pertaining to kin selection and related concepts[5] writes, "I continue to use the word 'social' in the broadest way, to cover all aspects of evolutionary change that deal with the tension between conflict and cooperation."

I don't read West-Eberhard's original definition as authorizing such a broad expansion. Social selection *sensu* Moore and Frank is any evolution resulting from natural selection in a social setting. Although this broad definition subsumes both sexual selection, and social selection in my sense, it smothers West-Eberhard's original claim that social selection rests on competition. Social selection in Moore and Frank's expansion is no longer a hypothesis, nor falsifiable, but merely *names* a body of research. I think we should retain social selection as a hypothesis rather than a name. Sexual selection and West-Eberhard's version of social selection both claim that what is happening in nature differs from what social selection in my sense claims is happening, as summarized in Tables 1 and 2.

Still another definition of social selection has been proposed in connection to the evolution of animal communication by Yoshinari Tanaka.[6]

3. M.J. West-Eberhard, "Sexual Selection, Social Competition, and Evolution.,"*Proceedings of the American Philosophical Society* 123 (1979): 222–234; M.J. West-Eberhard, "Sexual Selection, Social Competition, and Speciation," *Quarterly Review of Biology* 58 (1983): 155–183; M.J. West-Eberhard, "Sexual Selection, Competitive Communication and Species-specific Signals," in *Insect Communication*, ed. T. Lewis (New York: Academic Press, 1984), 283–324.

4. J.B. Wolf, E.D. Brodie III, and A.J. Moore, "Interacting Phenotypes and the Evolutionary Process. II. Selection Resulting from Social Interactions," *American Naturalist* 153 (1999): 254–266.

5. Steven Frank, "Social Selection," in *Evolutionary Genetics: Concepts and Case Studies*, eds. C.W. Fox and J.B. Wolf (Oxford: Oxford University Press, 2006), 350–363; cf. also Steven Frank, *Foundations of Social Evolution* (Princeton: Princeton University Press, 1998).

6. Yoshinari Tanaka, "Social Selection and the Evolution of Animal Signals," *Evolution* 50 (2): 512–523 (1996).

He writes that "the selective force that results from (nonsexual) social interactions is distinguishable from both natural selection and sexual selection. Several authors have stressed the importance of social selection in the evolution of animal signals (Crook[7], West-Eberhard[8], Tanaka[9]). I define social selection as the selective force that arises when a signal influences the fitness of signalers or both the fitness of signalers and receivers."

In summary, social selection *sensu* West-Eberhard extends the sexual-selection narrative to include competition within a social setting for resources other than mates. Social selection *sensu* Moore and Frank is evolution resulting from any social interaction. Social selection *sensu* Tanaka is natural selection mediated by the exchange of signals. And social selection *sensu* Roughgarden arises from managing the social infrastructure from which offspring are produced—it includes cooperation as much as competition and revolves around negotiation more than "winning." The objective of negotiation is to maximize the quantity of offspring reared, not genetic quality, with competition for mates being of secondary importance. Social selection in my formulation does not extend sexual selection as the West-Eberhard formulation does, it offers an alternative narrative.

The ambiguity of the phrase "social selection" has caused confusion, especially among evolutionary psychologists. For example, after dismissing my use of social selection as "idiosyncratic," Nesse adopts an everything-social definition.[10] He then offers a sexual-selection-like narrative for group formation in human evolution. Nesse writes, "Competition to be chosen as a social partner can, like competition to be chosen as

7. J.H. Crook, "Sexual Selection, Dimorphism and Social Organization in the Primates," in *Sexual Selection and the Descent of Man* (Chicago: University of Chicago Press, 1972), 231–281.

8. M.J. West-Eberhard, "Sexual Selection, Social Competition, and Evolution," *Amer Phil Soc* 123 (1979): 222–234.

9. Y. Tanaka, "The Evolution of Social Communication Systems in a Subdivided Population," *J Theor Biol* 149 (1991): 145–163.10 Randolph M. Nesse, "Runaway Social Selection for Displays of Partner Value and Altruism," *Biological Theory* 2 (2007): 143–155.

10. Randolph M. Nesse, "Runaway Social Selection for Displays of Partner Value and Altruism," *Biological Theory* 2 (2007): 143–156.

a mate, result in runaway selection that shapes extreme traits. People prefer partners who display valuable resources and bestow them selectively on close partners." So, natural selection for traits that promote being chosen as a partner leads to the evolution of "altruism, moral capacities, [and] empathy." But such traits represent the "endpoint of runaway social selection" associated with "substantial deleterious effects," including "vulnerability to mental disorders."

Of course, this sexual-selection-like narrative can't work for all the reasons we've seen before. Nesse's narrative would flounder on what might be called the "paradox of the saint"—selection for altruism, moral capacities, and empathy to be chosen as a partner would weed out genes that didn't express these traits, and everyone would quickly wind up being altruistic, moral, and empathetic. Moreover, the limited scope for detecting genetic variation in these qualities would make evolution through partner choice impossible. Any new version of a sexual-selection-like narrative crashes and burns for the same reasons as the original sexual-selection narrative does. To generate a suite of objections to any newly proposed sexual-selection-like narrative, just globally replace each occurrence of "male" and "female" in the original narrative with their new counterparts.

Nesse's work shows the logical slippery slope that attends the everything-social definition of social selection. Defining social selection broadly enough to make it automatically true doesn't mean that specific sexual-selection-like narratives are true too. But for sexual-selection advocates, social selection always boils down to sexual selection in some guise anyway, so that for them the logical danger is moot. The pattern of thought that sees sexual-selection-like narratives everywhere is addictive, and to use Richard Dawkins' terminology, is a near-religious mental virus, a meme, whose infection can only be cured with a healthy dose of the scientific method in which falsifiable alternative hypotheses are put the test.

As this book delves into controversies in evolutionary biology, I trust readers not to lose sight of the *facts* on which evolution biology is based. Those facts are that all multicellular life belongs to a common family tree and that biological species can change through time (unlike physical species, such as the elements of the periodic table). Everyone who has looked into the matter agrees with these two facts.

Even the "intelligent-design" proponents agree. Intelligent design, which is not identical to creationism but is politically co-mingled with it, initially claimed that structures like the eye were too complicated to be explained by Darwinian evolution. Its proponents said such structures must have been introduced abruptly by some creator, usually, but not necessarily, understood to be a god. Intelligent design had a long run, and the political clout of its activists even encouraged George W. Bush to say that "both sides," intelligent design and evolution, needed to be taught in high school science. But the gig is up.

Today, intelligent design proponents accept these two main facts of evolution—all life is related and species are not static. The most recent book[11] by a scientist from the intelligent design school, Michael Behe, contains many passages that explicitly endorse the old age of the earth, and the common ancestry that humans share with other animals, including primates. This latest book by Behe offers a mid-course correction to intelligent design thought and shifts its focus away from structures like the eye to whether mutations occur often enough and in the right order for drug resistance in malaria to evolve.[12] So, what began as a grand challenge to Darwinism has devolved into technical arguments about mutation rates.

I trust few readers will be troubled that we and other living things are one another's kin. Many of us have been dismayed to discover previously unknown brothers or sisters. We can't choose our parents or kin. Starfish, worms, and plants, even rose bushes and redwood trees, are our distant relatives, whether we like it or not. Rather than troubling, for me it's appealing to think that all of life is united into one body though membership in a common family tree.

Still, there are real challenges ahead for evolutionary biology, not the make-believe challenges dreamed up by right-wing think tanks pushing

11. Michael J.Behe, *The Edge of Evolution: The Search for the Limits of Darwinism* (New York: Free Press, 2007).

12. Joan Roughgarden, "A Matter of Mutation," review of *The Edge of Evolution*, by Michael Behe, *Christian Century*, 30 October 2008, 24–26.

intelligent design. Is natural selection really "nature red tooth and claw" survival of the fittest as Spencer claimed? Or is natural selection actually about success through cooperation? Or some mix of both, and in what proportions?

This book *is* about theory. It is not a book of fact, but a book of theories, a book of ideas. It's an attempt to figure out what's going on when evolution happens. We know that species reproduce sexually for example. How did this come about? What is it about reproducing sexually that can account for why most species reproduce sexually, although some don't? As we'll see, there are several ideas about this. This book is about what we don't know of evolution, not what we do know.

With all these theories, how do we find out which ones are right? We test them. We set the alternative hypotheses against one another, take data, and then make the call. At least that's what we should do. But scientists, like the home-field umpire, don't like calling their local-hero hypothesis out at the plate. My job is like the coach of the visiting team demanding that the umpiring be fair. I think the sexual-selection team has been getting a free pass around the baseball diamond, riding the coat tails of the correct parts of Darwinism. I think we need to call sexual selection out after its three strikes, and let the social-selection team get up to bat.

I can't overemphasize that evolutionary theories, like theories in any other area of biology, such as molecular biology, genetics, and physiology, are destined for testing. The reason for stating this hopefully obvious point is to counter the impression from evolutionary psychologists and the popular media that evolutionary theories are little more than me-Tarzan-you-Jane tall stories about our cave-man ancestors.

The Genetic System for Sex

FOUR

The Gene

RECOMBINATION

The three chapters of this part are devoted to why the genetic system that underlies sexual reproduction has evolved. The issues are: why has sex evolved, why have two gamete sizes evolved—small sperm and large eggs rather than gametes all the same size, and why have some species evolved to contain individuals who are separately male and female, whereas other species contain individuals who are jointly male and female? The three chapters of this part establish the foundation for later chapters about the behavior that takes place during sexual reproduction.

To begin, why does sex exist at all? To reproduce? Well, in lots of species reproduction is not sexual. In these species, reproduction takes place through budding or fragmentation. Put a cutting from a grape vine in the ground. It sprouts roots and *voilà*, a new plant springs forth. So why bother with sex when there are simpler ways of reproducing? And even species that can reproduce by budding or fragmentation sometimes reproduce sexually. Why bother? There must be something about sex that is important in itself—the value of sex must be about *how* to reproduce, not whether to reproduce. The question of why to reproduce sexually instead of asexually has long been a "big" question in evolutionary biology, attracting a who's-who of evolutionary biologists over many decades. Most biologists probably think the question is still unsolved, although in my opinion an adequate, if not completely satisfying answer is now available.

The early work from population geneticists explored the idea that a species that reproduces sexually could somehow evolve faster than a species using nonsexual techniques.[1] The way this would supposedly work is that a favorable mutation could occur at one genetic locus, say a locus for heat tolerance, and descendants of this individual would carry this beneficial mutation and prosper as the climate was heating up. Meanwhile, another favorable mutation could occur in a different individual at a different genetic locus, say one to protect against UV rays. The descendants of this individual would carry this beneficial allele and prosper as fewer clouds caused more direct exposure to the sun. When descendants of both these individuals mate, some of their offspring would carry both favorable mutations, and they would prosper especially well in a climate that is both hotter and has more exposure to the sun's direct rays. Thus, favorable mutations in different individuals can be brought together through sexual reproduction. This bringing together of mutations from different lineages is called between-locus genetic recombination.

In an asexual population, both favorable mutations would have to occur within the descendants from the same individual. This will indeed happen if one waits long enough. But, the long waiting time for a second mutation to occur in descendants of the individual where the first mutation occurred

---

1. R. A. Fisher, *The Genetical Theory of Natural Selection* (Oxford: Clarendon Press, 1930); H. J. Muller, "Some Genetic Aspects of Sex," *American Naturalist* 68 (1932): 118–138; James F. Crow and Motoo Kimura, "Evolution in Sexual and Asexual Populations," *American Naturalist* 99 (1965): 439–450.

was hypothesized to slow down the evolution of an asexually reproducing population relative to a sexual population. The asexual species would, therefore, be at a disadvantage relative to the sexual species.

According to this hypothesis, sexually reproducing species should, in geologic time, outlive species that reproduce asexually. And yes, it's true for the most part that lineages of sexually reproducing species are generally longer-lived in the fossil record than lineages of asexual species. Clonal species are evolutionary dead ends. On an evolutionary time scale, almost all clonal species are recently derived from sexual ancestors. On the family-tree of species, asexual species are twigs, not long branches.[2]

However, the promising idea that between-locus recombination is the source of the advantage to sexual reproduction hit the rocks in the 1970s and later. Although the genetic recombination in sex does bring favorable mutations together, it also splits them apart. The descendants of an individual with two favorable mutations, say for heat tolerance and for UV protection, might contain only one of these as a result of recombination with lines that lack these genes. Hence, between-locus recombination hurts as much it helps—the good and bad features of recombination more or less cancel out, leaving little net advantage if any. Mathematical models in population genetics showed that the time to fix a double favorable mutant in a sexual population is only marginally better, if at all, than in an asexual population.[3]

At the same time as the sexual-species-evolve-faster hypothesis was crashing on the rocks from research in population genetics, it was nonetheless being publicized throughout the rest of evolutionary biology

2. G. L. Stebbins Jr, *Variation and Evolution in Plants* (New York: Columbia University Press, 1950); O. P. Judson and B. Normark, "Ancient Asexual Scandals," *TREE* 11 (1996): 41–46; R. Butlin, "The Costs and Benefits of Sex: New Insights from Old Asexual Lineages," *Nature Reviews Genetics* 3 (2002): 311–317; Eugene A. Gladyshev, Matthew Meselson, and Irina R. Arkhipova1, "Massive Horizontal Gene Transfer in Bdelloid Rotifers," *Science* 320 (2008): 1210–1213.

3. S. Karlin, "Sex and Infinity: A Mathematical Analysis of the Advantages and Disadvantages of Genetic Recombination," in *The Mathematical Theory of the Dynamics of Biological Populations*, eds. M. S. B. and R. W. Hiorns (New York: Academic Press, 1973), 155–194; F. B. Christiansen et al., "Waiting With and Without Recombination: The Time to Production of a Double Mutant," *Theor Pop Biol* 53 (1998): 199–215.

by using an attractive metaphor taken from the story of *Alice in Wonderland*. The "Red-Queen Hypothesis" imagines that every species is involved in a never-ending evolutionary arms race and must continually evolve to stay even.[4] Because of having between-locus recombination, a sexually reproducing population can supposedly evolve faster than an asexual population, and thus be better at keeping up its arms race against the rest of nature.

The Red-Queen Hypothesis suffers from all the difficulties that surround the double-edged sword of recombination—recombination brings favorable mutations together, but also tears them apart after they have been brought together. On balance, recombination is a wash so far as the speed of evolution is concerned. Therefore, the Red-Queen Hypothesis has no theoretical foundation. Researchers have also pointed out that the ecology of a predator-prey interaction may culminate in peaceful coexistence, and not necessarily in an arms race.[5]

Nonetheless, the Red-Queen Hypothesis is continually mentioned in textbooks and courses in behavioral ecology in spite of being theoretically vacuous. No direct evidence shows that any asexual species has gone extinct, because it couldn't keep up in some arms race relative to sexual species. The only empirical study of the Red-Queen Hypothesis reports that lizards from an asexually reproducing species of geckos have more mites living on their scales than lizards from a comparable sexually reproducing species, but this study is careful to note the absence of any hard evidence that the mites are deleterious.[6] I suspect that the continued popularity of the Red-Queen Hypothesis rests on its world view of nature as an evolutionary treadmill, a view that meshes nicely with the sexual-selection world view of nature as selfish and competitive.

4. L. Van Valen, "A New Evolutionary Law," *Evol Theor* 1 (1973): 1–18.

5. Nils Chr Stenseth and John Maynard Smith, "Coevolution in Ecosystems: Red Queen Evolution or Stasis?" *Evolution* 38 (1984): 870–880; Ulf Diekmann, Paul Marrow, and Richard Law, "Evolutionary Cycling in Predator-Prey Interactions: Population Dynamics and the Red Queen," *J Theor Biol* 176 (1995): 91–102.

6. C. Moritz et al., "Parasite Loads in Parthenogenetic and Sexual Lizards (*Heteronotia Binoei*): Support for the Red Queen Hypothesis," *Proc R. Soc Lond B* 244 (1991): 145–149.

Another hypothesis for the evolution of sex that is held in high regard by sexual-selection proponents is called "Muller's Ratchet."[7] It focusses on getting rid of bad genes rather than on bringing good genes together, and is the mirror image of the Red-Queen Hypothesis.

Here's how the Ratchet works: "In a population which reproduces asexually, and in which back mutation does not occur, all offspring of an individual will carry all deleterious mutants on the genome of a parent, together with any new deleterious mutations. Thus, the number of deleterious mutations in an individual is at least as many as his parent carried... once a line has k mutants, it will never have fewer than k, and the Ratchet winds on inexorably, to the detriment of the population."[8]

Because a sexual population possesses recombination, it's possible for members of a line that has accumulated k mutants to mate with members of a different line, and some of their offspring might have fewer than k mutants. Thus, Muller's Ratchet envisions the progressive decline of asexual populations as mutations accumulate in them, whereas a sexual population would seem able to escape this bleak genetic destiny.

Again though, recombination proves to be a double-edged sword. Although recombination might allow some parents in a line with k accumulated mutations in it to produce offspring with less than k mutations, it will also cause the line to produce some offspring with more than k mutations. The net effect of the recombination is slight and depends on the initial statistical association between the mutant genes (called their "linkage disequilibrium") and on the functional interaction between the genes (called their "epistasis").[9] No strong case can be made theoretically for a benefit to sexual reproduction compared with asexual reproduction based on Muller's Ratchet.

7. H.J. Muller, "The Relation of Recombination to Mutational Advance," *Mutat Res* 1 (1964): 2–9; J. Felsenstein, "The Evolutionary Advantage of Recombination," *Genetics* 78 (1974): 737–756.

8. John Haigh, "The Accumulation of Deleterious Genes in a Population–Muller's Ratchet," *Theoretical Population Biology* 14 (1978): 251–267.

9. Michael Desai, Daniel Weissman, and Marcus W. Feldman, "Evolution Can Favor Antagonistic Epistasis," *Genetics* 177 (2007): 1–10.

Concerning direct evidence, a report in 1990 showed that Muller's Ratchet could take place in an RNA virus under laboratory conditions, with the recombing lines achieving better fitness than the asexual lines.[10] Although this study is suggestive, it does not account for the popularity of Muller's Ratchet among sexual-selection proponents. What's appealing about Muller's Ratchet in a sexual-selection context is its focus on bad genes. The irony is that if Muller's Ratchet did cause the evolution of sexual reproduction, this process of sexual reproduction would purge the gene pool of deleterious mutations prior to any possible further weeding by female choice, rendering sexual selection superfluous.

Furthermore, neither Muller's Ratchet nor the Red-Queen Hypothesis address the basic facts of the matter. What we need to explain is why asexual species keep popping up merely as twigs on the tree of life. Both Muller's Ratchet and the Red-Queen Hypothesis could explain why asexual lineages are twigs rather than long branches, but don't explain why they keep popping up. According to both the Red-Queen and Muller's Ratchet hypotheses, asexual reproduction has no advantage whatsoever and, therefore, all species should be sexual. But asexual species do exist and although asexual species are a minority of species, they aren't a rare, minuscule, and ignorable fraction of all diversity. Nor do these hypotheses account for the tendency of species who are able to reproduce both ways to engage in asexual reproduction when ecological conditions are constant such as in the middle of growing seasons, and to switch to sexual reproduction when environmental conditions are about to change at the end of growing seasons. To account for the association between mode of reproduction and ecological circumstance, we turn to a different approach altogether.

The British evolutionary biologist, John Maynard Smith, once posed the question: What use is sex?[11] The American evolutionary biologist, G. C. Williams[12] answered with John Bonner's[13] conclusion that "sex is a

10. Lin Chao, "Fitness of RNA Virus Decreased by Muller's Ratchet," *Nature* 348 (1990): 454–455.

11. John Maynard Smith, "What Use is Sex?" *Journal of Theoretical Biology* 30 (1971): 319–335.

12. George C. Williams, *Sex and Evolution* (Princeton: Princeton University Press, 1975), 3.

13. John T. Bonner, "The Relation of Spore Formation to Recombination," *American Naturalist* 92 (1958): 193–200.

parental adaptation to the likelihood of the offspring having to face changed or uncertain conditions." The basis for this consensus is that species having capabilities for both sexual and asexual reproduction typically reproduce sexually at times, such as the end of summer, when conditions are soon to change, and asexually otherwise. For example, aphids, tiny insects that live on garden plants, reproduce clonally at the beginning of the growing season, and then switch to sexual reproduction at the end of the season. Aphids benefit from fast reproduction when colonizing an empty rose bush, but return to sexual reproduction when anticipating a change of conditions at the end of the season.[14] In 1975, Williams[15] concluded "that the association between sexual reproduction and changed conditions... is adequately supported, even though... what is implied by changed conditions is not yet clearly specified." So, what *is* implied for the evolution of sex by changing conditions?

Compared with sexually reproducing species, asexual species can be thought of as "weeds"—species specialized for quick growth and fast dispersal like plants that locate and colonize new patches of ground when they open up. The common dandelion of North America is a clonal reproducer whose sexual ancestors live in Europe.[16] Weeds eventually give up their territory to other species who are poorer colonizers, but better over the long term.[17]

Animal species can be "weeds too." Many lizards are asexual. In Hawaii, all-female gecko species are locally abundant and widespread throughout the South Pacific.[18] More all-female species live in

14. George C. Williams and J. B. Mitton, "Why Reproduce Sexually?" *J Theoretical Biology* 39 (1973): 545–554.

15. George C. Williams, 1975. *Sex and Evolution.* (Princeton: Princeton University Press, 1975), 7.

16. O. Solbrig, "The Population Biology of Dandelions," *American Scientist* 59 (1971): 686–694.

17. D. Tilman, "Constraints and Tradeoffs: Toward a Predictive Theory of Competition and Succession," *Oikos* 58 (1990): 3–15.

18. R. Radtkey et al., "When Species Collide: The Origin and Spread of an Asexual Species of Gecko," *Proc R. Soc Lond B* 259 (1995): 145–152; K. Petren and T. Case, "Habitat Structure Determines Competition Intensity and Invasion Success in Gecko Lizards," *Proc Nat Acad Sci USA* 95 (1998): 11739–11744.

Texas, New Mexico, and Mexico—varieties of whiptail lizards.[19] The all-female species of whiptail lizards live along stream beds, and sexually reproducing relatives typically live up-slope from the streams extending into the adjacent woods or other vegetation.[20] Every major river drainage basin in southwestern North America is a site where an all-female whiptail lizard species has evolved. More than eight all-female species are described from this area.[21] Still more all-female species of lizards are found in the Caucasus Mountains of Armenia and along the Amazon River of Brazil. All-female fish occur, too.[22] Indeed, all-female animal species are found among most major groups of vertebrates.[23]

Moreover, some species have two kinds of females, those who don't mate when reproducing and those who do mate. Examples include grasshoppers, locusts, moths, mosquitoes, roaches, fruit flies and bees among insects, as well as turkeys and chickens.[24] Over 80% of fruit-fly species have at least some all-female reproducing individuals. Although the majority of females in these species reproduce through mating, selection in the laboratory increased the proportion of females not needing to mate by 60-fold, yielding a vigorous all-female strain.[25]

19. C. Cole, "Evolution of Parthenogenetic Species of Reptiles," in *Intersexuality in the Animal Kingdom*, ed. R. Reinboth (New York: Springer Verlag, 1975).

20. O. Cuellar, "On the Ecology of Coexistence in Parthenogenetic and Bisexual Lizards of the Genus *Cnemidophorus*," *American Zoologist* 19 (1979): 773–786.

21. O. Cuellar, "Animal Parthenogenesis," *Science* 197 (1977): 837–843; see also L. D. Densmore et al., "Mitochondrial-DNA Analysis and the Origin and Relative Age of Parthenogenetic Lizards (Genus *Cnemidophorus*). IV. Nine *Semilineatus*-group Unisexuals," *Evolution* 43 (1989): 969–983; A. Cullum, "Phenotypic Variability of Physiological Traits in Populations of Sexual and Asexual Whiptail Lizards (Genus *Cnemidophorus*)," *Evol Ecol Res* 2 (2000): 841–855.

22. R. C. Vrijenhoek, "The Evolution of Clonal Diversity in *Poeciliopsis*," in *Evolutionary Genetics of Fishes*, ed. B. J. Turner (New York: Plenum Press, 1984), 399–429.

23. See also R. C. Vrijenhoek et al., "A List of the Known Unisexual Vertebrates," in *Evolution and Ecology of Unisexual Vertebrates*, eds. R. M. Dawley and J. P. Bogart (Albany: New York State Museum, 1989), 19–23.

24. O. Cuellar, "On the Origin of Parthenogenesis in Vertebrates: The Cytogenetic Factors," *The American Naturalist* 108 (1974): 625–648.

25. H. Carson, "Selection for Parthenogenesis in *Drosophila mercatorium*," *Genetics* 55 (1967):157–171.

Thus, all-female species are well known among animals. The geckoes who colonized the South Pacific, and the whiptail lizards of stream beds in New Mexico also make sense as animal counterparts of weedy plants—species who are successful where dispersal is a premium and/or where the habitat is continually being disturbed.

So, to address the distribution and occurrence of sexual and asexual species, we need a hypothesis that somehow connects fluctuating environments with sexual *vs.* asexual reproduction. I first heard what I think is the answer during a seminar from W. D. Hamilton on the evolution of sex sometime in the early 1980s. He presented complicated computer simulations about how parasites could reduce the population size of a host species, and that through successive bust-boom cycles of host-parasite dynamics, a sexual population could grow faster over the long run than an asexual population.[26]

What I discerned from the simulations Hamilton offered was that a sexual host population with genes for resistance to various forms of the parasite was always keeping these genes available in the gene pool to continually re-evolve resistance to the parasite while each evolved with respect to one another during bust-boom cycles of abundance. The whole setup of the model was certainly complicated, but the idea seemed simple: an asexual population would be caught with too many of the wrong genes at some stages of the bust-boom cycle, whereas a sexual population would have a better balance of genetic diversity to meet each stage of the bust-boom cycle. So, I thought, and still do, that Hamilton had solved the problem for the evolution of sex by connecting environmental change, specifically, environmental change mediated by host-parasite bust-boom cycles, to sexual versus asexual mode of reproduction.

Later that year I was doing field work in the Caribbean on *Anolis* lizards with Steve Pacala and Andy Dobson, both of whom are now

---

26. William D. Hamilton, "Sex Versus Non-sex Versus Parasite," *Oikos* 35 (1980): 282–290; W. D. Hamilton, P. A. Henderson, and N. Moran, "Fluctuations of Environment and Coevolved Antagonist Polymorphism as Factors in the Maintenance of Sex," in *Natural Selection and Social Behavior: Recent Research and Theory*, eds. R. D. Alexander and D. W. Tinkle (New York: Chiron, 1981), 363–381.

professors at Princeton University. One of our projects was to determine the parasite load in lizards, and Andy Dobson's speciality was host-parasite ecology.[27] He had just come from Oxford where W. D. Hamilton was on the faculty and Dobson was familiar with Hamilton's work. During discussions in the evenings it became clear that people varied on how to interpret Hamilton's computer simulations, and that most felt they pertained specifically to how host-parasite co-evolution could account for the persistence of sexual versus asexual populations. In contrast, I felt that the host-parasite dynamics in Hamilton's simulations were merely a vehicle toward a more general point about the difference between how sexual and asexual populations respond to an environment that has recurrent back-and-forth fluctuations. So, after these discussions, there seemed a need to condense what I felt was Hamilton's insight into the simplest possible model for the idea, which led to a paper that appeared in 1991.[28]

Here's how the model works for what I term the *genetic-portfolio balancing hypothesis*. Imagine two species next to one another that are identical in all respects except that one reproduces sexually and the other asexually. For example, a lawn is about to be covered by either of two species of dandelions, the common dandelion we see all the time in North America, which reproduces asexually, compared to a sexually reproducing dandelion-like plant from Europe. Which one will wind up covering the lawn, the sexual species or the asexual species?

Imagine further that there is genetic variation in both species that can be represented as A_1A_1, A_1A_2, and A_2A_2, where A_1 and A_2 are the names we give to two alleles that happen to occur at a locus called "A". This locus could code for whether the leaves have a fuzzy silver look to them that reflects light. Accordingly, the A_1A_1, A_1A_2, and A_2A_2 plants have very fuzzy, medium fuzzy, and absorptive leaves, respectively. Each generation experiences a certain level of sun. When it's very sunny, A_1A_1 is best, when there's medium sun, A_1A_2 is best, and when it is overcast,

27. A.P. Dobson et al., "The Parasites of *Anolis* Lizards in the Northern Lesser Antilles," *Oecologia* 91 (1992): 110–117.

28. J. Roughgarden, "The Evolution of Sex," *The American Naturalist* 138 (1991): 934–953.

A_2A_2 is best. Then generation to generation, which genotype is best fluctuates independently of the others corresponding to how much the sun shines that generation.

Now, the only difference between the species is that in an asexually reproducing species, each type clones itself, that is, A_1A_1 individuals produce A_1A_1 offspring, A_1A_2 individuals produce A_1A_2, offspring and A_2A_2 individuals produce A_2A_2. In contrast, in the sexual population each type of plant mates with all the others and their offspring follow Mendel's laws. For example, when A_1A_2 mates with another A_1A_2, their offspring are ¼ A_1A_1, ½ A_1A_2, and ¼ A_2A_2, according to Mendel's laws. In addition, there is continually back and forth recurrent mutation between A_1 and A_2 so that neither can be lost completely. So, what happens?

Well, eventually the sexual species takes over the lawn, even though the asexual species can have a good long run as being the most common. The reason is interesting. It's only a matter of time until the asexual species is caught with all its genes in the wrong basket, so to speak. An asexual species comes to a time when nearly all the plants are A_1A_1 when what would be best is A_2A_2 and the species gets hammered, whereas the sexual species continually maintains more of a balance among the genetic types of plants, and never gets hammered as badly. The asexual species can spread faster than the sexual species for short bursts of time when all its genes are luckily just what is needed at that time, whereas the sexual species always pays a temporary cost for maintaining some individuals who are not best for the current situation.[29] An asexual population is like a natural get-rich-quick scheme and a sexual population like a naturally balanced mutual fund. Most people who invest their assets in get-rich-quick schemes retire to the breadlines, whereas retirees

29. The gene pool of an asexual population becomes overcommitted to particular genotypes after a run of environmental circumstances favoring those genotypes, and when the environment changes, the population suffers a crash, which lowers its geometric mean fitness below that of a corresponding sexual population. The geometric mean through time of the arithmetic mean fitness is higher in a random-mating diploid sexual population than in a corresponding asexual population. Long-term survival of a population depends on a high geometric mean fitness, and therefore the sexual population eventually outperforms a corresponding asexual population.

who still own their homes are more likely to have invested in balanced mutual funds.

The balanced genetic portfolio hypothesis (hereafter called simply the Portfolio hypothesis) has a conspicuously different philosophical flavor from the previous Red-Queen and Ratchet hypotheses. According to the Portfolio hypothesis, bad genes are negligible. Almost all genes are good sometime, somewhere. Sure, some genes may be down on their luck today, but not tomorrow. On this view of nature, it's stupid to select mates on the basis of who's got the best genes today. Anyone bothering to select mates on the basis of genes better have a perfect crystal ball to see the future. Nor is each species engaged in an eternal treadmill-like arms race with all the other species in nature. The task is simply for each to safely navigate the unpredictable seas of life.

Moreover, the Portfolio hypothesis does indeed address the connection between sex and "the likelihood of the offspring having to face changed or uncertain conditions" that the theory was originally called upon to do by John Bonner and G. C. Williams, whereas the Red Queen and Ratchet do not address the phenomenon they're supposed to explain. And, as we've also seen, the Red Queen and Ratchet are probably not true—they don't make theoretical sense and have little or no evidence to support them.

The main objection raised against the Portfolio hypothesis is not scientific but ideological, that the Portfolio hypothesis is "politically correct," because it demonstrates a long-term value to diversity. I've learned that anyone throwing around the criticism of "political correctness" is trying to legitimize their own bigotry. The idea that genetic diversity consists primarily of genes all of which are beneficial at various times and places may be appealing or repugnant to some. The idea that genetic diversity consists primarily of accumulated small defects may be appealing or repugnant to others. The question before us is, which is true? Which idea explains the persistence of sexual populations through geologic time and the recurrent popping up of asexual species with weedy characteristics? On the basis of the evidence, my call is that the Portfolio hypothesis is superior to either the Red-Queen or Ratchet hypotheses.

After my paper on the Portfolio hypothesis appeared, I spent a term on sabbatical in Merton College at Oxford. I recall speaking with Bob May (now Lord May), a leading theoretical ecologist at Oxford, about the model. Although not a precise quote, he essentially said, "Of course, a sexual species is 'buffered'" and mentioned that he had also obtained the result in previous work.[30] Bob May is right, I think. The word, "buffered," says it all. And I think many people have independently developed the Portfolio hypothesis in some way. But then I also spoke with W. D. Hamilton about it, feeling that I was merely developing what was his idea to begin with. Bill Hamilton wasn't so sure. He was concerned that the Portfolio hypothesis pertains to diploid species and raised the question of plant species like mosses in which the prominent phase is haploid with a brief, reduced diploid phase. I agree the rationale for sexual reproduction in such species is a problem and don't have a ready response. Thus, it isn't clear whether some version of the Portfolio hypothesis is really what Hamilton had in mind when I heard his seminar presenting computer simulations. But in any case, the Portfolio hypothesis does now exist formally as an alternative to the Red-Queen and Ratchet hypotheses and should be tested against them.

The Portfolio hypothesis is the first element to the social-selection theoretical system and is positioned within the system as an alternative to the Red-Queen and Ratchet hypotheses from the sexual-selection system. According to the Portfolio hypothesis, sexual reproduction owes its very existence to cooperation. Each clone within an asexual species has its private array of genes and stands or falls over the long term on the success of those genes. But each line of a sexual population continually exchanges genes with another, taking the good with the bad, and over the long run this sharing of genomes avoids extinction and leads to greater evolutionary success than the selfish genetic possessiveness of asexual clones.

---

30. Robert M. May and Roy Anderson, "Epidemiology and Genetics in the Coevolution of Parasites and Hosts," *Proceedings of the Royal Society London B* 219 (1983): 281–331.

Even bacteria swap genes back and forth—their gene pool might be thought of as a genetic commons, literally a "community pool" of genes that allow bacterial species to survive forever.[31]

No discussion of how sexual reproduction evolves would be complete without mentioning what biologists in recent decades have thought to be sex's main evolutionary stumbling block, "the cost of meiosis." The problem is simple. Suppose males make no contribution whatsoever, directly or indirectly, to the number of young that are successfully reared into the next generation. In this situation, the growth rate of a population is determined solely by female productivity. By assumption, the males are superfluous to the population's growth rate. If this assumption is true, then an asexual population consisting solely of females grows twice as fast as a population that is only 50% females with the remainder being males. According to this picture, having males around reduces the population growth rate by a factor of ½. This reduction in the overall population growth rate resulting from the free-loading males is called the "cost of meiosis."[32]

Alternatively, the cost of meiosis may be expressed in terms of kin selection: a female that can raise a fixed number of offspring leaves twice as many of her genes in the next generation by reproducing asexually rather than sexually.[33]

31. Cf. Bonnie L. Bassler, "How Bacteria Talk to Each Other: Regulation of Gene Expression by Quorum Sensing," *Current Opinion in Microbiology* 2 (1999): 582–587; Cynthia B.Whitchurch et al., "Extracellular DNA Required for Bacterial Biofilm Formation," *Science* 295 (2002): 1487; Søren Molin and Tim Tolker-Nielsen, "Gene Transfer Occurs with Enhanced Efficiency in Biofilms and Induces Enhanced Stabilization of the Biofilm Structure," *Current Opinion in Biotechnology* 14 (2003): 255–261; Robert G. Beiko, Timothy J. Harlow, and Mark A. Ragan, "Highways of Gene Sharing in Prokaryotes," *Proc Nat Acad Sci USA* 102 (2005): 14332–14337; W. Ford Doolittle and Eric Bapteste, "Pattern Pluralism and the Tree of Life Hypothesis," *Proc Nat Acad Sci USA* 104 (2007): 2043–2049; Davide Pisani, James A. Cotton, and James O. McInerney, "Supertrees Disentangle the Chimerical Origin of Eukaryotic Genomes," *Mol Biol Evol* 24 (2007): 1752–1760; John Bohannon, "Confusing Kinships," *Science* 320 (2008): 1031–1033.

32. John Maynard Smith, "What Use is Sex?" *Journal of Theoretical Biology* 30 (1971): 319–335.

33. Brian Charlesworth, "The Cost of Sex in Relation to Mating System," *Journal of Theoretical Biology* 84 (1980): 655–671; David Lloyd, "Benefits and Handicaps of Sexual Reproduction," *Evolutionary Biology* 13 (1980): 69–111; Marcy Uyenoyama, "On the Evolution of Parthenogenesis: a Genetic Representation of the 'Cost of Meiosis,'" *Evolution* 38 (1984): 87–102.

Because the cost of meiosis may be as high as a 50% loss of population growth rate, the dilemma is to find an advantage to sexual reproduction that offers some gain by a factor of two or more. The net advantage to sexual reproduction would then exceed the cost, and the evolution of sex will have been explained. But as Williams[34] writes, "Anyone familiar with accepted evolutionary thought will realize what an unlikely sort of quest this is. . . . Nothing remotely approaching an advantage that could balance the cost of meiosis has been suggested. The impossibility of sex. . . would seem to be. . . firmly established. . . . Yet this conclusion must surely be wrong."

Therefore, about half of the paper in which I developed the population-genetic model of the Portfolio hypothesis was devoted to analyzing whether the cost of meiosis is indeed a factor of ½ and comparing that cost to the benefit of sexual reproduction achieved through portfolio re-balancing. I confirmed that the cost need not be as high as ½ because its magnitude depends on whether females mate with a random sample of all males, at one extreme, or with solely one male at the other extreme, or with some mixture between these extremes. The cost is zero if females mate with only one male, and can be as high as ½ only if they mate at random, and lies between these endpoints for mixed mating arrangements.[35] Thus, my earlier analysis argued that the advantage to sexual reproduction from the genetic portfolio rebalancing would need to be weighed against the cost, which might be high, or low, depending on the mating arrangements.[36]

Today though, I'm no longer persuaded that the cost of meiosis is fundamental and would no longer bother with this cost-of-sex to ben-efit-of-sex comparison. It now strikes me that the cost of meiosis is

34. George C. Williams, *Sex and Evolution* (Princeton: Princeton University Press, 1975), 75.

35. E. L. Charnov, "Sex Allocation and Local Mate Competition in Barnacles," *Marine Biology Letters* 1 (1980): 269–272; Peter T. Raimondi and Janine E. Martin, "Evidence that Mating Group Size Affects Allocation of Reproductive Resources in a Simultaneous Hermaphrodite," *The American Naturalist* 138 (1991): 1206–1217.

36. The Red Queen and Ratchet hypotheses, by the way, never produce an advantage anywhere near as high as the Portfolio hypothesis can, and would never be able to counter a significant cost of meiosis.

simply an unjustified sexist assumption. Do males really start from a position of total indifference to how many young the females raise? This is the worst-case scenario of male parental care and is taken as the starting point for assessing the cost of meiosis. Why? The more eggs females produce, the more offspring can be sired. For males to walk out on their reproductive destiny with the females they've mated with makes no sense. Instead, complete male indifference to female productivity, if it exists, should be a special case arising only when there's nothing a male can possibly do to augment female productivity in a way that yields an increase in the number of young he sires that enters the next generation.

The explanation for why asexual species keep popping up and quickly dying compared with sexual species would seem to be completely explained by thinking of asexual species as genetic versions of get-rich-schemes and of sexual populations as genetic versions of long-term mutual funds, without any need to invoke cost-of-meiosis considerations.

FIVE The Cell

SPERM AND EGG

Every culture's creation myth offers a story for the origin of man and woman. A story familiar in Western culture resides in *Genesis*, Chapter 2 of the Bible, "And the rib, with the Lord God had taken from man, made he a woman, and brought her unto the man" (*Gen.* 2:22, KJV). According to this story, man and woman comprise a binary, and woman is derived from man.[1]

A story from Navajo tradition holds that the earliest people were "First Man" and "First Woman," who were created equally at the same

---

1. Other passages in the Bible acknowledge people in categories between men and women, and includes them within the religious communities of both the Hebrew and Christian Testaments: Isa. 56:4,5; Matt. 19:12; Acts 8:27–38. The Bible as a whole doesn't enforce a gender/sex binary, only this one famous passage from Genesis 2 does.

time.[2] They lived with Turquoise Boy and White Shell Girl who were "*nadle*," which means "changing one" or "one who is transformed" and refers to gender-variant "two-spirited" people. The two-spirited Turquoise Boy and White Shell Girl invented a pottery bowl, a plate, a water dipper, a pipe, a basket, axes and grinding stones from rocks, and hoes from bones. First Man and First Woman used these tools. People were dependent on the inventiveness of *nadle* who were present from the earliest eras of human existence as part of the natural order of the universe. In the Navajo tradition, neither sex is derived from the other; they have inherently equal status, and although male and female are categories, the two-spirited people are, too, which implies that the various sex/gender categories are not presented as binaries.

Biology also offers a creation story for male and female. Unlike religious myths, biology's creation stories are offered as scientific claims and are, therefore, subject to scientific examination. Even though ostensibly matters of science, biology's male/female creation stories have ideological overtones. These overtones pertain to whether sexual conflict is primitive and ubiquitous. Sexual-selection advocates claim that universal sexual conflict traces to the very origin of the male/female distinction. If sexual conflict is as old as the male/female distinction itself, sexual-selection advocates feel justified in taking sexual conflict as foundational to all social behavior.

The task before us once again is to overlook whether we find sexual-selection's creation story appealing or repugnant, but to determine so far as we can whether it is true. A first step is to develop alternative hypotheses to sexual-selection's story for the biological origin of male and female. And once again, we will see the sexual-conflict side try to preempt the scientific process by asserting that their position is obviously correct at the onset and it doesn't need to be tested or confronted with alternative hypotheses.

To begin, it's important to distinguish the concepts of "sex" from "male" and "female." Sex means producing offspring by combining the

---

2. Walter Williams, *The Spirit and the Flesh: Sexual Diversity in American Indian Culture* (Boston: Beacon Press, 1992), 19.

gametes from two parents. Having gametes means the reproduction is sexual and not having gametes means the reproduction is asexual. "Male" and "female" are another matter and pertain to the relative *size* of those gametes. "Male" means producing small gametes and "female" means producing large gametes. The male/female binary is defined in terms of gamete size. Species in which the gametes are the same size do not posses a male/female distinction even though they reproduce sexually.

To obtain a male/female distinction through evolution, two different gamete sizes must evolve from the presumably original condition of a single gamete size. Species in which the gametes have the same size are called "isogamous." Those with two gamete sizes are called "anisogamous." Isogamy is probably ancestral, because it is found only among primitive taxa of algae, fungi, and protozoa.[3] Hence, in biology, the evolution of male and female is filed under the heading of the "evolution of anisogamy," terminology that seems to bury the subject in biological jargon even though the origin of male and female is foundational to every culture's story of creation.

So what does evolutionary biology have to say about the evolution of male and female? The first models on the subject had nothing to do with sexual conflict, but pertained to how to bring the gametes into *contact* with one another. A dichotomy between contact versus conflict as the reason for why two gamete sizes evolved is the basic point of disagreement between the social-selection and sexual-selection hypotheses for the evolution of male and female.

In 1932, Kalmus focused on how zygotes are formed by contacts between eggs and sperm. The requirement for gametic contacts implies that the number of zygotes produced is proportional to the product of the number of eggs and number of sperm.[4] Provided the zygotes have to be bigger than some minimum size to survive, this product is maximized when one gamete is nearly the size of the zygote to begin with and the other gamete is as tiny as possible.

3. G. Bell, "The Evolution of Anisogamy," *J Theor Biol* 73 (1978): 247–270.

4. H. Kalmus, "Ueber den Erhaltungswert den Phaenotypischen Anisogamie und die Entstehung der Ersten Geschlectsunterschiede," *Biol Zentral* 52 (1932): 716.

Here's the intuition. A parent is trying, so to speak, to divide up the material it can place into eggs and sperm to maximize the number of gametic contacts that produce viable zygotes. If both proto-sperm and proto-egg are the same size, then only so many of them will bump into each other in the ocean's water where life began. The number of gametic contacts increases as gametes become more numerous, forming a more dense cloud of gametes. In principle, if both gametes could be made tiny, then when a dense cloud of tiny sperm mixes with another dense cloud of equally tiny eggs, the highest number of contacts will occur. But egg and sperm can't both be tiny and still produce zygotes big enough to survive. So, the maximum number of contacts producing viable zygotes then occurs when one of the gametes is nearly the desired zygote size while the other is as small as possible. A sparse cloud of large zygote-sized gametes colliding with a dense cloud of tiny gametes produces more contacts than do collisions between two medium-dense clouds each with medium-sized gametes. Thus, one possible advantage to gametic dimorphism in size is simply to maximize the number of contacts between gametes, subject to the constraint that the resulting zygotes must be big enough to survive. Kalmus' paper was later extended in 1967 by Scudo to incorporate the depletion of gametes as they underwent fertilization.[5]

This hypothesis was dismissed during the 1970s by Geoffrey Parker of the University of Liverpool. Parker argued that selection for anisogamy to maximize the contact rate required group selection and lacked a plausible explanation in terms of individual-level natural selection. As Parker explained in 1978, "Earliest analyses of the evolution of anisogamy (Kalmus,[6] Kalmus and Smith,[7] refined by Scudo[8]) were mainly dependent

5. F. Scudo, "The Adaptive Value of Sexual Dimorphism: I, Anisogamy," *Evolution* 21 (1967): 285–291.

6. H. Kalmus, "Ueber den Erhaltungswert den Phaenotypischen Anisogamie und die Entstehung der Ersten Geschlectsunterschiede," *Biol Zentral* 52 (1932): 716.

7. H. Kalmus and B. Smith, "Evolutionary Origin of Sexual Differentiation and the Sex-ratio," *Nature* 186 (1960): 1004–1006.

8. F. Scudo, "The Adaptive Value of Sexual Dimorphism: I, Anisogamy," *Evolution* 21 (1967): 285–291.

on group or population selection. Where there is a fixed amount of gametic reserve available in the population, and where gametes require a certain quantity of reserve for survival and development, these authors were able to show that the greatest number of successful fusions occurs when the gametic reserve is divided with a high degree of anisogamy than with isogamy."[9] In 1979, Parker further explained, "Though there are theories for the establishment of anisogamy through group or inter-population selection (Kalmus,[10] Kalmus and Smith,[11] Scudo,[12]), the 'anisogamy via disruptive selection' theory [Parker's name for his own theory] relies entirely on immediate forms of selection."[13]

Thus, as of the early 1970s, the contact-rate advantage to anisogamy was believed to require group selection. Indeed, the Kalmus and Scudo papers did demonstrate a population-growth-rate advantage to anisogamy. However, the fact that a trait confers a population advantage does not imply that it cannot also confer an individual advantage. Nonetheless, Parker and colleagues[14] introduced a model for the evolution of anisogamy that they claimed explicitly relied on individual selection, a model hereafter referred to as the PBS model.

The PBS model shares some features of the Kalmus model. Each individual releases gametes of one size into the gamete pool with a tradeoff between size and number—a few big gametes versus lots of small gametes. Because each individual makes only one size of gamete, it can be classified as male or female depending on whether it makes a small or large gamete, respectively.[15] However, PBS also allow any two gametes to fuse with each other to form a zygote whose volume is the sum of the

9. G. A. Parker, "Selection on Non-random Fusion of Gametes During the Evolution of Anisogamy," *J Theor Biol* 73 (1978): 1–28.

10. Kalmus, Ueber den Erhaltungswert," 716.

11. Kalmus and Smith, "Evolutionary Origin," 1004–1006.

12. Scudo, "Adaptive Value of Sexual Dimorphism," 285–291.

13. G. A. Parker, "Sexual Selection and Sexual Conflict," in *Sexual Selection and Reproductive Competition in Insects*, eds. M. S. Blum and N. A. Blum (New York: Academic Press, 1979), 123–166.

14. G. Parker, R. Baker, and V. Smith, "The Origin and Evolution of Gamete Dimorphism and the Male-Female Phenomenon," *J Theor Biol* 36 (1972): 529–553.

15. The next chapter examines in detail how sex should be "packaged" into bodies.

volumes of the component gametes. Thus, the PBS model allows zygotes to be made by sperm-sperm fusions, sperm-egg fusions, and egg-egg fusions. Hence, there are three possible zygote sizes based on two gamete sizes. PBS also assume that zygote survival increases as the zygote becomes bigger, so sperm-sperm zygotes survive the poorest, followed by sperm-egg zygotes, and egg-egg zygotes survive the best.[16]

The bottom line from the PBS model is that under some mathematical conditions, a polymorphism in gamete size evolves—the gamete pool does not wind up with only egg or only sperm, but with a mixture of both sizes. All-sperm leads to zygotes who can't survive very well, all-egg leads to just a few gametes that don't contact one another very often, and a sperm-egg mixture is better by comparison with either extreme. Sperm can enter a gamete pool with all eggs because of having a higher contact rate, and eggs can enter a gamete pool with all sperm because their zygotes survive better, resulting in a polymorphism of gamete sizes because each gamete size can increase when rare. Because an adult is assumed to make only one size of gamete, the sperm-egg dimorphism that evolves corresponds to the evolution of distinct males and females.

The theory today is framed more explicitly in terms of a competitive game between a proto-sperm and proto-egg that need to fuse with each other.[17] Imagine that initially both the proto-sperm and proto-egg are the same size. Then the proto-sperm "cheats," so to speak, by becoming a little smaller than the proto-egg so that more sperm can be produced with the leftover energy. A numerical advantage in sperm production then allows that size of sperm to out-compete the less numerous sperm with the original size. But a slightly smaller zygote now results, and it is less viable than the full-size zygote. Therefore, the proto-egg responds by increasing its size, restoring zygote viability to its original level. The egg's compensating in this way is better than becoming smaller itself to match the smaller sperm, because then the zygote would suffer a very deleteri-

16. They used a power-law formula to relate zygote size to survival, a detail relevant later when considering empirical tests.

17. Cf. H. Matsuda and P. Abrams, "Why are Equally Sized Gametes so Rare? The Instability of Isogamy and the Cost of Anisogamy," *Evol Ecol Res* 1 (1999): 769–784.

ous double loss of investment. A sequence of the sperm and egg's "best response" to each other then culminates in one gamete becoming nearly as large as the zygote, and the other gamete becoming as tiny as possible.

PBS added a verbal narrative to their mathematical model to draw out its meaning in terms of individual selection, a narrative that remains fundamental to the sexual-selection perspective today. PBS interpret the sperm-egg dimorphism as *conflict* between the gamete producers. PBS state, "Males are dependent on females and propagate at their expense, rather as in a parasite-host relationship." Parker reiterates this point in a later 1979 paper, "The primordial sexual conflict concerned the establishment of anisogamy itself. Ovum-producers would fare better without males to 'parasitize' the investment in each ovum. Males are likely to have won this earliest conflict...."[18]

According to the PBS model's narrative, the distinction between male and female is forged in battle: In the beginning, natural selection created male and female as enemies. Sexual-selection theory has then adopted the sexual conflict narrative from the PBS model. The "battle of the gametes" logically and temporally precedes the "battle of the sexes" that sexual-selection advocates take as the starting premise for theorizing about reproductive social behavior.

Because of the strategic significance of the male/female origin to the universality of sexual conflict, my laboratory began to review the anisogamy literature in the spring of 2006 during our weekly lab meeting. Meanwhile, Priya Iyer, a mathematics masters-degree graduate from the Indian Institute of Technology whose studies had included topics, such as Gödel's completeness theorem, regular graphs, and commutative algebra, joined my laboratory as a graduate student. She had become interested in combining ecological modeling with population genetics.[19] So, in the summer of 2006, prior to beginning her second year in graduate

18 G. Parker, "Sexual Selection and Sexual Conflict," in *Sexual Selection and Reproductive Competition in Insects*, eds. M. Blum and N. A. Blum (New York: Academic Press, 1979), 123–166.

19 J. Roughgarden, *Theory of Population Genetics and Evolutionary Ecology: An Introduction.* (New York: Macmillan, 1979; reprint, New York: Macmillan, 1987; reprint, Upper Saddle River, N.J.: Prentice Hall, 1996), 612.

school, we tried to see if there was something that needed to be done concerning the evolution of anisogamy.

The problem as I saw it was a disconnect between the PBS narrative and the PBS model—does the model actually entail the explanatory narrative offered on its behalf? The PBS journal article uses a combination of graphical argument and computer simulation, and mixes narrative with mathematics in a way that makes it difficult to tell exactly how the derivation is unfolding. Our lab group couldn't discern anything wrong mathematically with the PBS model, but wasn't convinced the analysis was either correct or that the model actually implied the narrative.

Our lab group's uncertainty about how to evaluate the PBS model with its attendant narrative reinforced my own skepticism about the role of sexual conflict in the origin of male and female. If male and female were created by natural selection as enemies, then shouldn't species revert to asexual reproduction whenever the conflict becomes sufficiently intense? The sexual-conflict narrative undercuts any of the theories, whether Portfolio, Red-Queen, or Ratchet, for why sex evolves to begin with, because if the deleterious effect of sexual conflict were severe enough, then the population cost of sexual conflict would exceed the population advantage to sexual reproduction, and the sexual population would revert to asexual reproduction.

As we continued reading the anisogamy literature, we encountered a 1978 paper by Charlesworth[20] that clarified matters considerably and became the starting point for the research we eventually carried out. Charlesworth proposed a population-genetic model for a hypothetical genetic locus that determines gamete size.[21] The adults were haploid, so that an individual with, say, allele A_1 produces many tiny gametes and an individual with allele A_2 produces a few large gametes. These gametes might be thought of as proto-sperm and proto-eggs and the adults who made them could, therefore, be considered as males and females, respectively. The zygote's size was the sum of the gamete sizes who fused to

---

20. B. Charlesworth, "The Population Genetics of Anisogamy," *J Theor Biol* 73 (1978): 347–357.

21. Cf. Charlesworth, "Population Genetics," 347–349.

make it.[22] From this model Charlesworth writes, "The conclusions of Parker[23] based on computer calculations, are confirmed analytically." Specifically, Charlesworth derived the condition for a polymorphism to be maintained between small-gametes and large-gametes instead of an all-sperm or all-egg gamete pool.[24]

I found the Charlesworth formulation especially interesting for two reasons. First, nowhere in Charlesworth's paper does the phrase, "sexual conflict," appear, nor even the word, "conflict." Nowhere is there any reference to sperm as "parasitizing" the egg. Thus, the sexual-conflict narrative of the PBS paper is gratuitous, a story attached to the PBS model as though the model required it.

Second is the mathematical technique Charlesworth used to confirm PBS model conclusions. He showed that the PBS model could be rearranged to a version of "the standard equation for gene frequency change in a single-locus multiallele system in a random mating diploid." To readers unfamiliar with population genetic models, this is the equation for how a gene pool changes as a result of natural selection, an equation found in all elementary evolution textbooks. Charlesworth showed how the PBS model could be framed in terms of the constant fitness coefficients used in the standard textbook equation by following this recipe: fitness equals number of collisions times the zygote survival.[25] To Charlesworth, the connection between the PBS model and the familiar one-locus model in population genetics was primarily instrumental. He writes, "We can therefore use the well-known properties of this system to investigate [the] equation." No interpretation was offered for the meaning behind the recipe for the fitness coefficients.

To me though, the recipe for the fitness coefficients was the most significant aspect of the single-locus formulation because the recipe indicates

22. Like PBS, the probability of a zygote's survival was assumed to depend on its size according to a power law.

23. G. Parker, R. Baker, and V. Smith, "The Origin and Evolution of Gamete Dimorphism and the Male-Female Phenomenon," *J Theor Biol* 36 (1972): 529–553.

24. The condition is that the curve relating zygote size to survival must increase much faster than a straight line, as it does in a power-law formula.

25. Cf. Charlesworth, "Population Genetics," equation (3a).

what's happening as evolution proceeds. In a single-locus system with constant fitness coefficients, the population-average fitness increases each generation, culminating at a point where average fitness is maximized—a result is called the "Fundamental Theorem of Natural Selection" in population genetics. So, the recipe tells us what is being maximized through time by natural selection. Well, what's being maximized is precisely the contact rate leading to viable zygotes, that is, the product of the egg and sperm numbers, times the probability of survival. So, the PBS model actually shows how important contact rate, qualified by survival probability, is to evolving two gamete sizes. The original PBS model is not about gametic conflict after all, but about gametic contact.

Priya and I then decided to follow up the Charlesworth model with our own formulation to highlight and make more explicit the role of contact instead of conflict in the evolution of anisogamy, hereafter called the IR model.[26] We considered the evolution of anisogamy in individuals who are able to make both eggs and sperm, such as plants and hermaphroditic marine invertebrates. We postulated a hypothetical locus that simultaneously controls both the sperm size and egg size produced by an individual. Also, we allowed the hermaphrodite adults to be diploid. Thus in our model, the evolution of anisogamy involves fixing an allele that codes for extreme size differences between the two gamete types made by an individual. In our model, anisogamy does not represent a genetic polymorphism, but a gene pool that is genetically monomorphic for a gene that codes for a certain phenotypic expression, namely the production by an individual of gametes with extreme size differences. In this model, gametic conflict is inherently impossible, because the adaptive value of the anisogamy among the gametes produced by an individual is credited to that individual. Anisogamy is, thus, a trait by which an individual increases its own fitness according to ordinary individual-based natural selection.

The model's development parallels the Charlesworth treatment in that the gene-pool dynamics also turns out to be governed by the familiar textbook equation for gene-frequency change at one locus, and also turns out to use a recipe for constant fitness coefficients based on multiplying

26. Priya Iyer and Joan Roughgarden, "Gametic Conflict Versus Contact in the Evolution of Anisogamy," *Theor Popul Biol* 73 (2008): 461–472.

the contact rate (number of eggs times number of sperm) with the survival probability. One difference with earlier work is that we used a curve relating zygote survival to size that has previously been employed in studies of the evolution of larval life spans for marine invertebrates, a curve called the Vance survival function[27] instead of the power law curve used by Charlesworth and PBS.

The conclusions from the IR model are first, that the population evolves isogamy, and hence no male/female distinction, when an individual faces little risk in making both its gamete types tiny, even though the resulting zygotes are tiny too, because doing so maximizes contact rate. This situation occurs when the survival probability for zygotes is high, even if they are small, and/or if the developmental growth rate of individual zygotes is high. Second, the population evolves anisogamy with its corresponding male/female distinction when an individual faces great risk in making both its gametes small. This situation occurs survival probability for small zygotes is low and/or the developmental growth rate of individual zygotes is slow. In this situation, making many tiny zygotes that fail to survive during the time needed to attain the size at which they transform into juveniles is useless effort. Instead, contact rate leading to zygotes that actually survive is maximized when one gamete is nearly the desired zygote size to begin with and the other gamete is tiny. Therefore, a testable prediction of the IR model is that both gametes in an isogamous species should be tiny and sperm-sized, whereas in an anisogamous species, one gamete should be tiny and sperm-sized and the other large and nearly final-zygote-sized.

Priya then examined the gamete sizes, initial and final zygote sizes, and development time for isogamous and anisogamous species of *Volvocaceae*, a family of colonial green algae.[28] The zygote (zygospore) growth rate was calculated as (final zygote size–initial zygote size)/development time.

27. R. Vance, "On Reproductive Strategies in Marine Benthic Invertebrates," *Am Nat* 107 (1973): 339–352. See also: J. Roughgarden, "The Evolution of Marine Life Cycles," in *Mathematical Evolutionary Theory*, ed. M. W. Feldman (Princeton: Princeton University Press, 1988), 270–300.

28. R. Levine and W. Ebersold, "The Genetics and Cytology of *Chlamydomonas*," *Annu Rev Microbiol.* 14 (1960): 197–216.

As predicted from the IR model, gametes of isogamous *Volvocaceae* species are statistically significantly smaller than the eggs of anisogamous species and not significantly different in size from the sperm. As also predicted, isogamous *Volvocaceae* species have much smaller zygotes compared to anisogamous species. Furthermore, the percent increase in zygote size during development to the mature zygospore stage is significantly higher in isogamous species than in anisogamous species, confirming the model prediction that isogamous zygotes have higher zygotic growth rates.

Meanwhile, how has the PBS model stacked up against the data? Unfortunately, the tests so far have simply accepted the PBS narrative as correct, and instead have sought empirical support for the details assumed in PBS about how zygote survival relates to size. For example, Randerson and Hurst[29] discuss the alga, *Chlamydomonas*, which has same-sized gametes in two self-incompatible varieties, plus and minus. They accept the PBS narrative to begin with by writing, "Imagine a mutant plus type producing multiple small gametes. Such a mutant can, therefore, fertilize numerous proto-eggs, whilst contributing little to the zygote. These proto-sperm parasitize the investment provided by proto-eggs. Although it is easy to see why such a 'cheating' strategy should evolve, it is less obvious how the original large gamete strategy could be maintained in the face of such exploitation. Why should minus types not also produce large numbers of small gametes?" They then provide PBS's own answer to this question: "PBS requires that zygote fitness must increase disproportionately with volume (i.e., increments in zygote size must confer more than proportional increments in fitness), at least over part of its size range. This assumption is by no means standard, and we are aware of no direct supportive evidence." Their paper then reviews the many attempts to reconcile the PBS model with data on the shape of the curve relating zygote survival to size. I'm not aware of any empirical studies that zoom in on the central conceptual issue before us, whether anisogamy is results from gametic conflict, or promotes gametic contact.

The topic of how anisogamy evolves has by now accumulated a modest literature, although not nearly as large as that for the evolution of

29. James P. Randerson and Laurence D. Hurst, "The uncertain evolution of the sexes," *TRENDS in Ecology and Evolution* 16 (2001): 571–579.

sex.[30] Improving the assumptions about how zygote survival relates to zygote size and refining the description of gametic-contact kinetics to include features like the relative target size and pheromone releasing capacity of small versus large eggs is valuable. Yet, such improvements can distract from the central issue—whether contact versus conflict is the cause of anisogamy's evolution. Of the papers that support a gametic-contact position, some propose an advantage in group productivity or species competition. The continuing taint of group selection discourages many evolutionists from considering gametic-contact hypotheses. Such evolutionists feel that only the gametic-conflict hypothesis is true to the spirit of individual-level selection. As we've seen though, this sentiment is a mistake. Contrary to the PBS narrative about gametic conflict, gametic *contact* is precisely what is being favored by individual-level natural selection in both the original PBS model itself, as explicated by Charlesworth, as well as in our own IR model.

I anticipate some readers are now gasping after having read more than they ever wanted to know about the sex life of sea weeds. Lest one think the evolution of anisogamy is nothing but the obsessive preoccupation of microbiology wonks, let me return to what's at stake here, why we should all care about the sex life of these tiny green plants.

On the first page of their recent monograph, *Sexual Conflict*,[31] Göran Arnqvist and Locke Rowe offer putative examples of sexual conflict in

30. N. Knowlton, "A Note on the Evolution of Gamete Dimorphism," *J Theor Biol* 46 (1974): 283–285; J. Madsen and D. Waller, A Note on the Evolution of Gamete Dimorphism in Algae," *Am Nat* 121 (1983): 443–447; R. F. Hoekstra, "Evolution of Gamete Motility Differences. II: Interaction with the Evolution of Anisogamy," *J Theor Biol* 107 (1984): 71–83; R. Hoekstra, "The Evolution of Sexes," in *The Evolution of.Sex and its Consequences*, ed. S. Stearns (Basel: Birkhäuser, 1987), 59–92; D. Levitan, "Effects of Gamete Traits on Fertilization in the Sea and the Evolution of Sexual Ddimorphism," *Nature* 382 (1996): 153–155; D. Levitan, "Optimal Egg Size in Marine Invertebrates: Theory and Phylogenetic Analysis of the Critical Relationship Between Egg Size and Development Time in Echinoids," *Am Nat* 156 (2000): 175–192; D. Dusenbery, "Selection for High Gamete Encounter Rates Explains the Success of Male and Female Mating Types," *J Theor Biol* 202 (2000): 1–10; D. B. Dusenbery, "Selection for High Gamete Encounter Rates Explains the Evolution of Anisogamy Using Plausible Assumptions About Size Relationships of Swimming Speed and Duration," *J Theor Biol* 241 (2006): 33–38; T. Togashi, P. A. Cox, and J. L. Bartelt, "Underwater Fertilization Dynamics of Marine Green Algae," *Math Biosci* 209 (2007): 205–221.

31. Göran Arnqvist and Locke Rowe, *Sexual Conflict* (Princeton: Princeton University Press, 2005).

robber flies, penduline tits, funnel-web spiders, and the Malabar ricefish. Of the male ricefish, they write, "he strikes her in the genital region with a complex clublike organ. If the organ, a modified anal fin, contacts the female body, it releases a spermatophore (a sperm capsule) with a dart-like spike. This spike is pushed into the female flesh, and the spermatophore becomes firmly attached because of a whorl of recurved barbs at its tip. Females are adapted to these repeated assaults." They then pose the question, "what do we make of these? It is difficult to reconcile observations of open conflict between the sexes with the common view of mating as a joint venture of two individuals that, by virtue of being the same species, share a common genome." Arnqvist and Locke's answer to their question is that, "despite interacting males and females sharing the same genome, conflict between them is ubiquitous."

Is sexual conflict ubiquitous? One might reply to Arnqvist and Locke that their descriptions seem anthropomorphic and that many species do not exhibit behavior that could possibly be interpreted as open male-female conflict. But no, to Arnqvist and Locke the matter is already settled. They write, "Conflict exists because there are two sexes, and therefore will be present in all anisogamous species, and has neither an evolutionary starting point nor an end."[32] They continue, "The term 'resolution,' or 'resolved,' is often used in context of sexual conflict… It is important to note that resolution does not in any real sense make sexual conflict disappear or even fade. There is no solution to sexual conflict that somehow causes the evolutionary interests of individuals of the two sexes to coincide."[33]

Yet, what is the basis for these confident assertions that sexual conflict is universal?—the very theories we have just examined whose tests so far are confined to the sex life of tiny sea weeds. The PBS narrative for the evolution of anisogamy anchors the ideology of universal sexual conflict in nature. But as we have seen, the PBS narrative that sperm and egg result from gametic conflict has no theoretical basis and no empirical support. The alternative narrative of anisogamy as a trait promoting gametic contact does have both theoretical basis and empirical support.

32. Ibid., 218.
33. Ibid., 219–220.

Göran Arnqvist is an editor of the journal, *Evolution*, with responsibility for manuscripts concerning sexual selection and sexual-conflict. Other journals too provide editorial support for sexual conflict. *The American Naturalist* devoted a supplemental issue to sexual conflict.[34] The *Philosophical Transactions of the Royal Society*, Series B, featured a collection of papers suggesting that sexual conflict might merit the status of a new "paradigm" in evolutionary biology.[35] Its papers were drawn from a meeting in 2005 with over 200 people who met at the Royal Society in London to discuss sexual conflict. The organizers state, "Our motivation for organizing the meeting was that it seemed to us that in the last ten years or so, there has been a dramatic shift in the prevailing view of matings as being essentially 'a good thing' for both participants, to one in which they are regarded as 'bad' for females." Instead of being a new paradigm, sexual conflict is a fad.

Sexual conflict is rape in scientific guise, a narrative of males victimizing females.[36] One may find this brave new world championed by the sexual-conflict advocates appealing or repugnant. But is it true and accurate? As has been shown, its anchor in the evolution of anisogamy is vacuous.

Instead, according to the social-selection position, natural selection created male and female as allies, not enemies. The male/female distinction was forged in cooperation, not battle.

From a parent's perspective, making anisogamous gametes is to its benefit because of the higher contact rates those gametes achieve.

From an egg's perspective, it benefits from the presence of sperm and is not parasitized by them. If sperm weren't tiny, an egg would languish unfertilized until desiccation and death. An egg may trade some material it would acquire from fusing with a sperm as large as itself in return for a much greater chance of being fertilized when sperm are tiny. The rationale for such a trade assumes that fertilization increases with the number of sperm, a relation termed *sperm limitation* as contrasted with *sperm saturation* in which increasing sperm does not affect whether eggs are fertilized, but

34. David Hosken and Rhonda Snook, "How Important is Sexual Conflict?" *Am Nat* 165 (2005): S1–S4.

35. T. Tregenza, N. Wedell, and T. Chapman, "Introduction. Sexual Conflict: A New Paradigm?" *Phil Trans R. Soc B* 361 (2006): 229–234.

36. R. A. Johnstone and L. Keller, "How Males Can Gain by Harming Their Mates: Sexual Conflict, Seminal Toxins, and the Cost of Mating," *Am Nat* 156 (2000): 368–377.

only affects who does the fertilizing. Sperm limitations occur in the marine environment among free-spawning species because of low gamete density.[37]

The sperm's view does allow for some degree of competition however. Small sperm are more numerous than large sperm and therefore might seem advantaged in sperm competition. But sperm in the same ejaculate are related to one another, facilitating the evolution through kin selection of sperm cooperation. Sperm may cooperate to reach the eggs using coordinated swimming assisted by elaborate rowboat-like structures.[38] So, even sperm are not necessarily condemned to eternal struggle against each other.

In summary, the contact hypothesis for the evolution of anisogamy is the second element in the social-selection theoretical system and is alternative to the conflict hypothesis found in the sexual-selection system. In social selection, the production of sperm and eggs is a trait whereby parents attain the highest number of fertilized zygotes rather than, as sexual selection postulates, the outcome of a miniature battle between the sexes being played out among microscopic gamete cells.

The creation story for male and female in evolutionary biology proceeds through stages. At the beginning is the evolution of sexual reproduction and in the middle is the evolution of two gamete sizes. Then we arrive at the stage whereby whole individuals come to contain male and/or female function. I term this third stage, *sexual packaging*—it's about the types of sexual function an individual contains within its body. Does an organism produce *both* eggs and sperm? If so, in what ratios? Or, does an individual produce *only* eggs or *only* sperm? And across the entire population, what types of bodies occur—solely sperm-producers and egg-producers, solely body types that make both eggs and sperm, or various mixtures of all of these? The packaging of sexual function into bodies is the next logical stage of evolutionary biology's creation story for male and female.

37. D.R. Levitan and C. Petersen, "Sperm Limitation in the Sea," *Trends Ecol Evol* 10 (1995): 228–291; D. Levitan, "Effects of Gamete Traits on Fertilization in the Sea and the Evolution of Sexual Dimorphism," *Nature* 382 (1996): 153–155.

38. Harry Moore et al., "Exceptional Sperm Cooperation in the Wood Mouse," *Nature* 418 (2002): 174–177; F. Hayashi, "Sperm Cooperation in the Fishfly, *Parachauliodes Japonicus*," *Functional Ecology* 12 (1998): 347–350; J. Sivinski, "Sperm in Competition," in *Sperm Competition and the Evolution of Animal Mating Systems*, ed. R.L. Smith (New York: Academic Press, 1984), 223–249.

SIX

The Body

MALE, FEMALE, AND HERMAPHRODITE

If a single body makes only sperm during its life span, it is a "male," and if only eggs, it is a "female." If a single body produces both eggs and sperm during its life span, it is called a "hermaphrodite." If both egg and sperm are produced at the same time, the body is called a "simultaneous hermaphrodite," or if at different and possibly overlapping times, the body is a "sequential hermaphrodite."

Here are the issues. First, the distribution of bodily packaging arrangements has not been clear—do most species consist of males and females, do most species consist of hermaphrodites, or do most species contain all three body plans at the same time? Second, which came first? Did separate male and female bodies coalesce into hermaphrodite bodies or did hermaphrodite bodies disarticulate into separate male and female bodies? The sexual-selection position has always taken distinct male and female bodies as the initial condition and viewed hermaphrodites as derived from them. The social-selection position is the opposite—ancient hermaphrodite lineages have given rise to separate male and female bodies as a specialization. Third, what is the behavior of hermaphrodites? Is a hermaphrodite a strangely troubled creature with its male half at war with its female half? The sexual-selection position sees sexual conflict everywhere, even between the two halves of the common body of a hermaphrodite. The social selection position views a hermaphrodite as an integrated body whose production of two sizes of gametes maximizes the number of its zygotes that survive to enter the next generation. In this chapter, we'll take up these issues one by one.

A male/female binary at the whole-body level is nowhere close to being universal across the plant and animal kingdoms. Hermaphroditism is the rule in plants, and only about 6% of species have separate males and females out of 250,000 total species.[1] The pattern is just the opposite among animals.[2] Separate males and females is the rule in animals, and only about 5% to 6% of species are hermaphroditic out of over 1.2 million total species. The number of animal species that are hermaphroditic amounts to about 65,000. The proportion of hermaphrodites among animals becomes much higher once insects are removed from the total because insects are not hermaphroditic and yet insects comprise well over half, about a million, of all animal species, leaving us with a figure of ⅓ hermaphrodite species among all animal species excluding insects.

All in all, across all the plants and animals combined, the number of species that are hermaphroditic is more-or-less tied with the number

1. S.S. Renner and R.E. Ricklefs, "Dioecy and its Correlates in the Flowering Plants," *Am J Bot* 82 (1995): 596–606; D.W. Vogler and S. Kalisz, "Sex Among the Flowers: The Distribution of Plant Mating Systems," *Evolution* 55 (2001): 202–204; J.C. Vamosi and S.M. Vamosi, "The Role of Diversification in Causing the Correlates of Dioecy," *Evolution* 58 (2004): 723–731; C. Goodwillie, S. Kalisz, and C.G. Eckert, "The Evolutionary Enigma of Mixed Mating Systems in Plants: Occurrence, Theoretical Explanations, and Empirical Evidence," *Annu Rev Ecol Syst* 36 (2005): 47–79.

2. Philippe Jarne and Josh R. Auld, "Animals Mix it up too: The Distribution of Self-fertilization Among Hermaphroditic Animals," *Evolution* 60 (2006): 1816–1824.

who has separate males and females, and neither arrangement of sexual packaging can be viewed as the "norm." In this situation, it is not obvious which bodily arrangement to take as the starting point. Do separate males and females descend from hermaphrodites or *vice versa*? Which body plan to take as the starting point is a major difference between the sexual-selection and social-selection positions concerning the evolution of sexual packaging.

Available empirical information could support both views as to which body plan to take as the starting point. Depending on the group of organisms being discussed, evolutionary transitions can seemingly go in either direction. Moreover, whether hermaphroditic bodies or separately-sexed bodies last for a long time during the evolutionary record varies too.[3]

Hermaphrodism is stable over evolutionary time in trematodes (flukes, a class of flatworms parasitic on mollusks and vertebrates)[4] and the pulmonate gastropods (land and freshwater snails with a special lung), although these groups are very old, pre-Cambrian in age. Among these pulmonate gastropods, at least 40 evolutionary shifts to hermaphroditism from the ancestral separately sexed body state have occurred. Conversely, evolutionary shifts from hermaphrodism to separately sexed bodies and then back again have happened in other groups, such as the Cnidarians (corals, jellyfish, hydra) and the prosobranch gastropods (most marine snails).[5] The barnacles, too, a group that I have worked on, have done some crisscrossing in their sexual packaging. The common acorn and gooseneck barnacles found in the rocky intertidal zone are hermaphroditic and descended from shrimp-like ancestors

3. In the biological literature, the condition of separately sexed bodies is called dioecy or gonochorism.

4. Only the schistosomes have evolved separately sexed bodies from hermaphroditic ancestors, cf. Thomas R. Platt and Daniel R. Brooks, "Evolution of the Schistosomes (Digenea: Schistosomatoidea): The Origin of Dioecy and Colonization of the Venous System," *J Parasitology* 83 (6), 1035–1044 (1997).

5. Philippe Jarne and Josh R. Auld "Animals Mix it up too: The Distribution of Self-fertilization Among Hermaphroditic Animals," *Evolution* 60 (2006): 1816–1824; J. Heller, "Hermaphroditism in Mollusks," *Biol J Linn Soc* 48 (1993): 19–42.

who had separately sexed bodies. Yet, these hermaphroditic barnacles in turn have been ancestral to some barnacles with separately sexed bodies that are parasites upon large marine animals, such as whales and turtles.[6]

Nor is hermaphroditism restricted to invertebrates and plants. Even our vertebrate relatives include many hermaphroditic species, especially among coral-reef fish. Most species of wrasses, parrotfishes, and larger groupers are hermaphroditic, as are some damselfish, angelfish, gobies, porgies, emperors, soapfishes, dottybacks, and moray eels—all from shallow waters, and many deep-sea fish as well.[7]

Sequential hermaphrodism manifests in several patterns. In the overwhelming majority of groups, the sex-changing individuals begin as male and transition into female (MtF).[8] Only two groups, the urochordates and vertebrates (fish) have any species in which some individuals transition from female to male (FtM).[9]

Among sequential hermaphrodite fish specifically, the MtF transitions typical of the invertebrate sequence occur in the clownfish, the colorful fish who live in sea anemones.[10] However, the atypical reverse transition,

6. J.T. Høeg, "Sex and the Single Cirripede: A Phylogenetic Perspective," in *New Frontiers in Barnacle Evolution*, eds. Frederick R. Schram and Jens T. Høeg (Rotterdam: A.A. Balkema, 1995), 195–207.

7. R. Warner, "Mating Behavior and Hermaphrodism in Coral Reef Fishes," *American Scientist* 72 (1984): 128–136; Mead, G., E. Bertelson and D.M. Cohen, "Reproduction Among Deep-sea Fishes," *Deep Sea Research* 11 (1964): 569–596.

8. In the biological literature, this pattern is called "protandry," that is, "male-first." This pattern is recorded in the Cnidaria, Entoprocta, Cestoda, Bivalvia, Sipuncula, Polychaeta, Ectoprocta, Gastrotricha, Malacostraca, Echinodermata, Urochordata, Vertebrata (fish); cf. Offline Appendix 2 to Philippe Jarne and Josh R. Auld, "Animals Mix it up too: The Distribution of Self-fertilization Among Hermaphroditic Animals," *Evolution* 60 (2005): 1815–1824.

9. This sequence is called "protogyny", that is, "female first."

10. H. Fricke and S. Fricke, "Monogamy and Sex Change by Aggressive Dominance in Coral Reef Fish," *Nature* 255 (1977): 830–832; J. Moyer and A. Nakazono, "Protandrous Hermaphrodism in Six Species of the Amenonefish Genus *Amphiprion* in Japan," *Japan J Ichthyology* 25 (1978): 101–105; P.M. Buston, "Size and Growth Modification in Clownfish," *Nature* 424 (2003): 145–145; P.M. Buston and M.A. Cant, "A New Perspective on Size Hierarchies in Nature: Patterns, Causes, and Consequences," *Oecologia* 149 (2005): 352–372; P.M. Buston and M.B. García, "An Extraordinary Life Span Estimate for the Clown Anemonefish *Amphiprion Percula. J Fish Biology* 70 (2007): 1710–1719.

FtM, occurs in many fish species, including the well-studied blue-headed wrasse.[11] Still other fish species have crisscross sex changing. In one species of goby, some individuals mature from an unsexed juvenile to a female, then transition to a male, and then back to a female again (FtMtF).[12] In another species of goby, some individuals mature from a juvenile to a male, then transition to a female, and then back to a male again (MtFtM).[13] Bidirectional sex changes in gobies are receiving increasing study.[14] Nor do these various sex-changing scenarios exhaust the possibilities. Some coral reef fish, especially the sea basses, do not change sex at all but exist as simultaneous hermaphrodites.[15]

Finally, patterns of sexual packaging also include what are called "dwarf males." These tiny males are mobile testes. Among the fish, the deep sea angler fish offer a spectrum of examples. In some species, angler fish males are not only tiny, but are often incapable of a free-living

11. D. Robertson, "Social Control of Sex Reversal in a Coral Reef Fish," *Science* 1977 (1972): 1007–1009; D. Robertson and R. Warner, "Sexual Patterns in the Labroid Fishes of the Western Caribbean, II: The Parrotfishes (Scaridae)," *Smithsonian Contributions Zoology* 255 (1978): 1–25; R. Warner and S. Hoffman, "Local Population Size as a Determinant of a Mating System and Sexual Composition in Two Tropical Reef Fishes (*Thalassoma* spp.)," *Evolution* 34 (1980): 508–518; J. Godwin, D. Crews, and R. Warner, "Behavioral Sex Change in the Absence of Gonads in a Coral Reef Fish," *Roc R. Soc Lond B* 253 (1995): 1583–1588. See also: J. Moyer and A. Nakazono, "Population Structure, Reproductive Behavior and Protogynous Hermaphrodism in the Angelfish *Centropyge Interruptus* at Miyake-jima, Japan," *Japan J Ichthyology* 25 (1978): 25–39.

12. T. Kuamura, Y. Nakashima, and Y. Yogo, "Sex Change in Either Direction by Growth-rate Advantage in the Monogamous Coral Goby, *Paragobiodon echinocephalus*," *Behavioral Ecology* 5 (1994): 434–438.

13. P. Munday, M. Caley, and G. Jones, "Bi-directional Sex Change in a Coral-dwelling Goby," *Behav Ecol Sociobiol* 43 (1998): 371–377.

14. P.L. Munday et al., "Diversity and Flexibility of Sex Change Strategies in Animals," *Trends Ecology Evolution* 21 (2005): 89–95; E.W. Rodgers, R.L. Earley, and M.S. Grober, "Social Status Determines Sexual Phenotype in the Bi-directional Sex Changing Bluebanded Goby *Lythrypnus dalli*," *J Fish Biology* 70 (2007): 1550–1558.

15. C.W. Petersen, "Reproductive Behaviour and Gender Allocation in *Serranus Fasciatus*, a Hermaphroditic Reef Fish," *Anim Behav* 35 (1987): 1501–1514; C.W. Petersen and E.A. Fischer, "Mating System of the Hermaphroditic Coral-reef Fish, *Serranus baldwini*," *Behav Ecol Sociobiol* 19 (1985): 171–178; C.W. Petersen, "Reproductive Behavior, Egg Trading, and Correlates of Male Mating Success in the Simultaneous Hermaphrodite, *Serranus tabacarius*," *Env Biol Fish* 43 (1995): 351–351; Christopher W. Petersen, "Sexual- Selection and Reproductive Success in Hermaphroditic Seabasses," *Integrative Comparative Biology* 45 (2005): 439–448.

existence. In such species, the males have nostrils for homing in on pheromones released by females. The males also have pinchers to grasp projections on the female. After a male attaches to the back or side of a female, their epidermal tissues fuse and circulatory systems unite.[16] Although some species have attaching dwarf males that fuse with the body of a female as just described, other species have a polymorphism of both free-living males and attaching males, and still other species have males who are exclusively free-living. Over 100 species of anglerfish live throughout the world at depths below one mile, including one in the relatively accessible deep trench of Monterey Bay, California.[17]

Dwarf males are also typical of spiders, and occur in numerous crustacean species including the barnacles as well as some marine snails and rotifers.[18] Multiple tiny males are usually associated with each female, constituting polyandry. I was once able to handle a female angler fish with attached males that had been preserved in alcohol and had been collected during a deep sea oceanographic expedition. It felt like a golf ball with bumps—the bumps were the males and the golf ball was the female. In some species, like the barnacles, the dwarf males occur together with hermaphrodites, and in other species, like the angler fish, the dwarf males occur together with females.

The phenomenon of dwarf males might seem esoteric, but actually raises a basic question about sexual packaging. Why aren't all males dwarf males? If all that males ever provide during reproduction is sperm, then why aren't all males heat-seeking ballistic testes? A sexual-selection type of answer might be that males must compete with each other to acquire access to females, and therefore must be large enough to win at male-male combat. But then any size to a male in excess of that needed for testes would have to have some function in male-male combat. This simply isn't true. Male bodies are more than testes with guns.

16. E. Bertelsen, "The Ceratioid Fishes. Ontogeny, Taxonomy, Distribution, and Biology," *Dana Report* 39 (1951): 276 pp.; T. Pietsch, "Dimorphism, Parasitism and Sex: Reproductive Strategies Among Deep Sea Ceratioid Anglerfishes," *Copeia* 1975: 781–793.

17. T. Fast, "The Occurrence of the Deep-sea Anglerfish, *Cryptopsaras Couesii* in Monterey Bay, California," *Copeia* 1957: 237–240.

18. Fritz Vollrath, "Dwarf Males," *TREE* 13 (1998): 159–153.

Alternatively, males might in general supply much more to reproduction than merely their genes. Although exactly what they contribute to successful offspring rearing might not be easy to discern in the field, males should be doing something that increases the number of offspring successfully raised, because otherwise they should all be dwarf sperm-sacs like male spiders and male angler fish.

Anyway, if you happened to think that the male/female binary is inscribed in biological nature, forget it. At the gamete level, the binary in gamete size is nearly universal, but that's it, so far as universal generalizations about male and female go. At the whole-body level, the diversity of sexual packages in nature is seemingly bewildering and is decidedly not a universal binary. This diversity implies that a theory for the evolution of sexual packaging will have to predict a diversity of outcomes under various conditions, and the task ahead for evolutionary theory on this issue is challenging.

So, let's see if we can figure out how sexual packaging might have evolved, focussing initially on simultaneous hermaphrodites, and then moving to sequential hermaphrodites.

Sexual-selection advocates take the position that the whole-organism male/female binary is primitive and hermaphrodism is derived. For example, the concluding sentence of the PBS paper in 1972 states, "Hermaphrodism has probably evolved several times independently from the primitive situation of two separate sexes."[19] The evolution of sexual packaging is, therefore, framed in terms of seeking reasons why a hermaphroditic body plan could evolve from a separate-sex body plan in special circumstances. Simultaneous and sequential hermaphrodism are considered distinct phenomena in both social and sexual selection.

The advantage to hermaphrodism is not about self fertilization. Hermaphrodites usually mate with one another, not with themselves, and the distribution of self-compatibility is not the issue.[20]

---

19. G. Parker, R. Baker, and V. Smith, "The Origin and Evolution of Gamete Dimorphism and the Male-Female Phenomenon," *J Theor Biol* 35 (1972): 529–553.

20. Philippe Jarne and Josh R. Auld, "Animals Mix it up too: The Distribution of Self-fertilization Among Hermaphroditic Animals," *Evolution* 50 (2005): 1815–1824.

To explain simultaneous hermaphrodism, sexual-selection advocates rely on an early paper by Tomlinson[21] that introduces what might be termed the *dating game* advantage for hermaphrodites versus separately sexed organisms: the task of finding a compatible partner.

Tomlinson argues that hermaphrodism has an "increasing advantage as populations become sparse, when the populations are thinly scattered in temporary or marginal habitats, and when the effective breeding area is reduced by lack of motility as in sessile or sluggish animals without widely distributed gametes." Here's Tomlinson's reason: "If the chance of encountering another animal is extremely small, it is highly advantageous to have that rare individual of a type that would insure fertilizing capability. In hermaphrodites this rarely-encountered individual will always have fertilizing capabilities, while the gonochoristic [separately sexed] individual's chances are one-half (assuming an equal sex ratio). Therefore, the chances of at least one successful contact favors the hermaphrodite by a factor of at least two."

Tomlinson is implicitly envisioning organisms who mate by internal fertilization, not by broadcasting their gametes into open water. A pair wise encounter between two individuals would not be necessary in broadcast fertilizers.

Yes, any two internally fertilizing hermaphrodites who mate can produce gametes that fuse, whereas internally fertilizing separately sexed individuals must await a male-female encounter to yield gametes that fuse. However, as Heath pointed out in 1977, "hermaphrodites have nothing but advantages when compared to gonochorists [separately sexed bodies]."[22] Tomlinson's model shows that for internal fertilizers, hermaphrodism is *absolutely* better than separately sexed bodies regardless of population density, even though the size of this benefit, but not its sign, does depend on population density. Therefore, the sexual-selection position requires some unspecified additional reason to account for why

21. J.T. Tomlinson, "The Advantages of Hermaphroditism and Parthenogenesis," *J Theor Biol* 11 (1955): 54–58.

22. D.J. Heath, "Simultaneous Hermaphroditism: Cost and Benefit," *J Theor Biol* 54 (1977): 353–373.

all internal fertilizing species are not hermaphroditic. Moreover, the sexual-selection position focusses on hermaphrodites as a specialization to extremely low-density circumstances, which is clearly does not apply to the vast majority of simultaneously hermaphroditic species.

In contrast, the social selection position is that separate sexes have evolved as a specialization from an initially hermaphroditic body plan. The social-selection modeling, therefore, goes in the opposite direction from the sexual-selection modeling—what has to be explained according to sexual selection is why hermaphrodites exist, whereas what has to be explained according to social selection is why separate sexes exist.

Priya Iyer, whose research on the evolution of anisogamy has already been mentioned, has now published a model for how the evolutionary process might go from hermaphrodism to separate sexes.[23] The idea behind the model is that splitting male and female function into separate bodies is a specialization whereby males provide "home delivery" of sperm to the vicinity of eggs. In theory, males may evolve within a population that consists initially solely of hermaphrodites, because they possess a higher efficiency at sperm delivery, an efficiency achieved at the cost of a lower ability to produce eggs. According to this theory, after such males have come to exist in the population together with the original hermaphrodites, the remaining hermaphrodites then evolve to reallocate their gonadal investment solely into eggs because the sperm they had been producing are now superfluous. This scenario leads to distinct males and females from an original hermaphroditic state, provided that males attain a more efficient "home delivery" of sperm than a hermaphrodite.

Whether males indeed provide more efficient sperm delivery than a hermaphrodite in turn depends on how well they perform in local ecological circumstances. In conditions of low abundance where animals are sparse and separated by large distances relative to their mobility, males

---

23. Priya Iyer and Joan Roughgarden, "Dioecy as a Specialization Promoting Sperm Delivery," *Evolutionary Ecology Research* 10 (2008): 867–892.

may not be able to deliver sperm more effectively than hermaphrodites after allowance is made for their up to twofold dating-game disadvantage. In such circumstances, the initial hermaphroditic state would remain evolutionarily stable, or if separate sexes previously evolved in some other location or time, the species will evolutionarily revert to hermaphrodism.

Priya developed a mathematical model to ascertain the conditions for which the evolutionary transition from hermaphrodite to separate sexes can occur. Priya envisioned that "home delivery" by males boils down to supposing that their sperm are present in the vicinity of eggs in a higher concentration than if broadcast into the open water. By permitting egg-sperm contacts to take place in relatively confined volumes compared with open water, a higher contact rate is realized which increases the fertilization efficiency of sperm. Therefore, a gene for this capability can increase in a population initially consisting solely of hermaphrodites, thereby kick-starting the scenario leading to distinct males and females.

According to this scenario, the evolution of separate sexes from hermaphrodism is linked to a transition from broadcast, open spawning to a more localized spawning, including in the extreme, internal fertilization. Priya has gone on to investigate the association between hermaphrodism and fertilization mode across the family tree of animal phyla.[24] Priya collected data in all animal phyla about whether the sexual packaging consists of hermaphrodism or separate sexes. She also collected data in all animal phyla about whether the sperm are (a) broadcast— taken to include sperm released by many individuals or the entire population that are more-or-less uniformly distributed over eggs irrespective of whether the eggs themselves are released or fertilized inside the parent or (b) localized—taken to include internal fertilization, pseudo-copulation, hypodermic impregnation, and most cases of spermatophore release wherein the sperm are concentrated at the sites where the eggs are available. When diversity within a phylum in the

24. Kenneth M. Halanych, "The New View of Animal Phylogeny," *Annu Rev Ecol Evol Syst* 35 (2004): 229–255.

sexual packaging or fertilization characteristics was encountered, data at the various subphylum levels down to the family level were collected. All the data were mapped to family trees for the animal kingdom resolved to the level of phyla. Ancestral traits were reconstructed for hermaphrodism/separate-sex and localized/broadcast-spawning transitions at the branch points in the family tree.

Priya's family-tree analysis suggests that hermaphrodism and broadcast fertilizations are primitive among multicellular animal phyla, and that both separate sexes and localized fertilization are derived. However, back-transitions to hermaphrodism and broadcast fertilization also occur. Hermaphrodism in mollusks, annelids, arthropods and vertebrates is derived. Overall though, the family tree of animal phyla shows that hermaphrodism is in general primitive in the animal kingdom, and separate sexes derived. Furthermore, broadcast spawning is also generally primitive in the animal kingdom, and localized fertilization derived.

That both separate sexes and localized fertilization are jointly derived from hermaphrodism and broadcast fertilization offers broad-brush support for our model. Nonetheless, this support is tentative because the ancestral states reconstructed at many branch points in the family tree of phyla are ambiguous, and particular transitions to separate sexes may occur independently of transitions between localized and broadcast fertilizations, suggesting that the coupling between separate sexes and localized fertilization is not as tight as our scenario envisions.

We have not focused much on sequential hermaphrodism, as such, because we're presently uncertain about how to frame the phenomenon. If the transition from male to female or vice versa is rapid, then the population at any time effectively consists of separate sexes. In this case, the phenomenon of hermaphrodism poses a question more for evolutionary developmental biology (evo-devo) than for the evolution of social behavior. Conversely, if the transition in sex is slow, then the population consists at any one time of a polymorphism among simultaneous hermaphrodites, males and females. The evolution of this polymorphic situation is then subsumed under the modeling discussed above that pertains to the stages in how separate sexes may be derived from simultaneous hermaphrodism.

In contrast with the social selection position, sequential hermaphrodites are a big deal for sexual selection advocates. They introduce an argument for its evolution called the "size advantage" model. The idea is simple, animals can produce one type of gamete better while small, and another when big. This idea would seem to accord with the overwhelming majority of sequential hermaphrodites that transition from male to female as they age. The minimum body size that can produce sperm obviously is smaller than the minimum body size that can produce eggs, so the sex-change sequence should typically go from male to female (MtF) as the organisms age, which is indeed true.

Next, we get to the particular case of FtM sequential hermaphrodism in coral-reef fish. At this point a sexual-selection narrative explicitly arises, as championed by Ghiselin who declares in a recent autobiographical retrospective, "Darwin's theory of sexual selection was one of his most brilliant accomplishments and perhaps the one that has been the least well understood."[25] He writes that Darwin's *"The Descent of Man, and Selection in Relation to Sex* deserves to be read as a profound contribution to the philosophy of our subject. Understanding it that way certainly helped my own creativity." Ghiselin also states that, "it was my paper entitled 'The evolution of hermaphroditism among animals'[26] that first explained protandry and protogyny in terms of Darwin's theory of sexual selection."

In fact, sexual selection has nothing to do with protandric (MtF) hermaphrodism. Sexual selection might be relevant only to protogynous (FtM) hermaphrodism. That is, the size advantage hypothesized for MtF hermaphrodism is simply that small bodies make sperm and large bodies make eggs, which has nothing to do with sexual selection. Only the reverse direction, FtM, postulates a sexual-selection-like advantage for male size, as we now see. Moreover, crisscrossing sex changes (MtFtM and FtMtF) are not consistent with any sort of size-advantage model

25. Michael T. Ghiselin, "Sexual Selection in Hermaphrodites: Where Did Our Ideas Come From?" *Integrative Comparative Biology* 45 (2005): 358–372.

26. M. T. Ghiselin, "The Evolution of Hermaphroditism Among Animals," *Q Rev Biol* 44 (1959): 189–201.

unless the animals' body size expands and shrinks back and forth across some hypothetical threshold, which seems far fetched, or the threshold moves back and forth, so animals must continually change sex to keep up, which also seems far fetched. So, how is sexual selection supposedly relevant to FtM hermaphrodites?

Ghiselin continues to write that, "one day in the old biology library I read a paper on a sex-switching fish. It remarked that the males are brightly colored. At once I realized that sexual selection might be operative and had a classic 'aha' experience. It was not, however, a matter of the females choosing the more spectacular looking or aesthetically pleasing males, but rather of 'male combat' with males fighting and the successful contestants monopolizing the opportunities to fertilize the eggs. The fish reproduced as females until they were big enough to win fights, and then turned into males." This hypothesis is known as "harem polygyny"—females change sex and become males to control a harem of females. He adds that, "the hypothesis was remarkably original. Not only has nobody ever challenged my priority for the size advantage model, but no real adumbrations of it have also thus far turned up in the publications of any other scientist." Ghiselin concludes that "the size advantage model provides an excellent example of a successful hypothesis. After more than 3 decades of research, it has never had a serious competitor and there is no reason for abandoning it. In other words, it is probably true."

Well, is it true that the size advantage, if any, of FtM hermaphrodites is to control a harem by "winning fights"? Probably not.

Consider instead the bluebanded goby.[27] In this species two females can be placed together and one then transitions into a male. The reason for the transition can't be to win fights with other males—there are no other males to fight with. Instead, the sex change increases the reproductive success of the fish who switched, with the remaining female producing fertilized eggs and the newly minted male siring those eggs.

27. Edmund Rodgers, Shelia Drane, and Matthew Grober, "Sex Reversal in Pairs of *Lythrypnus Dalli*: Behavioral and Morphological Changes," *Biol Bull* 208 (2005): 120–125.

This observation suggests a different approach to understanding FtM fish and their "harems." Instead of sex changes to "win fights," let us hypothesize that each sex change takes place by a female when she can attain more sires than she can produce eggs, and thereby generalize the bluebanded goby situation. Then calling the collection of females present as a "harem" emerges as a mistake.

Suppose a female can make 10 eggs. Then two females → 0 fertilized eggs. Therefore, one female transitions, so that one female + one male → 10 fertilized eggs. This is the bluebanded goby situation where the female produces 10 fertilized eggs and the male realizes 10 sires. Moving to larger groups, two females + one male → 20 fertilized eggs, because each female gets 10 fertilized eggs apiece and the male gets 20 sires. Suppose a female transitions in this group. Then one female + two males → 10 eggs, because the remaining female gets 10 fertilized eggs and each male gets five sires. So, no female should transition. Next suppose there are three females and one male. Then each female gets 10 fertilized eggs and the male gets 30 sires. But if one of the females transitions, then there are two females and two males. Now each female gets 10 eggs and each male gets 10 sires, which is a wash from the female standpoint. So, this system will remain as three females plus one male, which sexual-selection advocates would incorrectly describe as a harem. The existence of the male would have nothing to do with winning fights against other males to control a harem of females. In this case, it's not numerically advantageous from a female perspective to bother changing sex. Finally, if there are four females and one male, and if one female transitions to male, we have three females and two males—each female will get 10 eggs and each male will get 15 sires. Thus, the possibility of male-male combat doesn't arise until the group is big enough that it would pay a female to transition even though a male is already on the scene.

Overall then, the sexual-selection narrative is irrelevant for most cases of sequential hermaphrodism, and even in the FtM case it is undemonstrated and probably incorrect.

Let us now take the existence of hermaphrodites as a given, and ask whether their behavior in some way exemplifies sexual selection. Of course, sexual-selection advocates say yes, and I say no.

The lead article by Leonard[28] to a recent symposium on sexual selection and hermaphrodites confidently begins, "Over the last 130 years, research has established that (a) sexual selection exists and is widespread in the plant and animal kingdoms; (b) it does not necessarily entail sexual dimorphism; even hermaphrodites have it; (c) it does not require intelligence or a sophisticated sense of esthetics; even tapeworms and plants choose mates; and (d) it does not require brawn or even mobility for competition; plants may compete for pollinators, and broadcast spawning invertebrates may also compete for matings." The article also asserts, "Sexual selection has come to be seen as a keystone of Charles Darwin's theory of evolution by natural selection," and goes on to say how sexual selection is now understood to be more general than even Darwin himself imagined: "Darwin considered sexual selection to be limited to higher animals (from arthropods on up) on two grounds; first, 'it is almost certain that these animals have too imperfect senses and much too low mental powers to feel mutual rivalry, or to appreciate each other's beauty or other attractions,'[29] and 'In the lowest classes the two sexes are not rarely united in the same individual, and therefore secondary sexual characters cannot be developed.'[30] Darwin, then, saw hermaphroditism as incompatible with sexual selection both because of a lack of opportunity for evolution of sexual dimorphism and a lack of capacity for mate choice and/or direct competition for mates in many invertebrates."

What could explain such confidence that sexual selection has become settled science as the "keystone" of natural selection, a status exceeding even Darwin's own assessment of its limitations? This is definition slippage.

To sexual selection advocates, sexual selection is true by definition. Leonard writes that Darwin, "defined sexual selection as depending 'on the advantage which certain individuals have over the same sex and

28. Janet L. Leonard, "Sexual Selection: Lessons from Hermaphrodite Mating Systems," *Integrative Comparative Biology* 45 (2005): 349–357.

29. C. Darwin, *The Descent of Man, and Selection in Relation to Sex* (Princeton: Princeton University Press, 1871), 321.

30. Darwin, *Descent of Man*, 321.

species, in exclusive relation to reproduction.'[31] As Andersson[32] has pointed out, this definition can be applied to all organisms." Correct, on this definition, sexual selection simply labels the traits whose adaptive value lies in reproduction rather than survival.

But then we see that "reproduction" quickly morphs into "mating," and "advantage" morphs into "competition." Leonard adopts for herself the following definition of sexual selection: "Selection through competition to acquire mates or be chosen as a mate." She writes, "The definition of sexual selection adopted here . . . was put forward by Malte Andersson . . . based on Darwin's definition of sexual selection in *Descent of Man* (quoted earlier). The key feature of the definition is competition; not just competition for access to mates, but competition to be chosen as a mate, and competition to choose the best (right) mates."

Now Leonard's definition is quite different from Darwin's. Sexual selection in this specific sense no longer automatically applies to all species, and alternatives such as social selection are possible. Replacing one definition that is automatically true with a second definition that may not be true in hopes of inheriting the former's generality is the fallacy of "definition slippage."

The drumbeat of praise defending sexual selection infects the research with confirmation bias. If sexual selection is assumed true at the onset, then research devolves into mining for behaviors that can be accreted to the sexual selection narrative. Darwin did underestimate the sophistication of animals "below" the arthropods. Yet, the ability of such animals to exhibit sexual selection is not testimony to their capabilities. If sexual selection is generally false in arthropods and above, it is also false for animals below the arthropods.

Readers may be wondering why they should care about the sex life of snails, slugs, and flukes that Leonard discusses. The reason is familiar. Claiming that the sex life of these creatures obeys the dictates of sexual selection is to argue that sexual conflict is universal through out nature—even slugs are mean and nasty. If you think wife beating is bad, wait to

---

31. Darwin, *Descent of Man*, 255.
32. M. Andersson, *Sexual Selection* (Princeton: Princeton University Press, 1994).

you see what slugs do to one another. Once again, the issue before us is not whether the prospect of slug wife beating is appealing or repugnant, the issue is whether sexual conflict truly is present in hermaphrodites, and possibly more severe there than in separate-sex species.

Sexual-selection advocates have sensationalized the mating behaviors of simultaneous hermaphrodites. Leonard[33] calls attention to "apophallation" in the giant amber-colored banana slugs of coastal redwood forests in California wherein copulation is followed occasionally by amputation of the penis of one or both individuals (penis biting).[34] In a different genus, individuals amputate their own penis after copulating and present it to their partner. Another behavior called "hypodermic insemination" involves injecting sperm under the partner's skin and is widespread among hermaphroditic groups, such as leeches, polyclad flatworms, and sea slugs, whereas only a few cases are known among separate-sex species, such as the bedbug.[35] "Penis fencing" in a species of marine flatworm with hypodermic insemination supposedly represents individuals attempting to ward off damaging penis insertions.[36] Likewise, hormone transfer during mating,[37] which also occurs in separate-sex species,[38] results from shooting "love darts" in the garden snail[39] and from body piercing with setae in earthworms.[40]

33. Janet L. Leonard, "Sexual Selection: Lessons from Hermaphrodite Mating Systems," *Integrative Comparative Biology* 45 (2005): 349–357.

34. H. Reise and JMC Hutchinson, "Penis-biting Slugs: Wild Claims and Confusions," *Trends Ecol Evol* 17 (2002): 153; H. Heath, "The Conjugation of *Ariolimax Californicus*," *Nautilus* 30 (1915): 22–24; A.R. Mead, "The Taxonomy, Biology, and Genital Physiology of the Giant West Coast Land Slugs of the Genus *Ariolimax Mörch* (Gastropoda: Pulmonata)" (Ph.D. diss., Cornell University, 1942).

35. A.D. Stutt and M.T. Siva-Jothy, "Traumatic Insemination and Sexual Conflict in the Bed Bug *Cimex Lectularius*," *Proc Natl Acad Sci USA* 98 (2001): 5583–5587

36. N.K. Michiels and L.J. Newman, "Sex and Violence in Hermaphrodites," *Nature* 391 (1998): 547.

37. J.M. Koene and A. Ter Maat, "'Allohormones': A Class of Bioactive Substances Favoured by Sexual Selection," *J Comp Physiol A* 187 (2001): 323–325.

38. C. Gillott, "Male Accessory Gland Secretions: Modulators of Female Reproductive Physiology and Behavior," *Annu Rev Entomol* 48 (2002): 153–184.

39. J.M. Koene and R. Chase, "Changes in the Reproductive System of the Snail *Helix Aspersa* Caused by Mucus From the Love Dart," *J Exp Biol* 201 (1998): 2313–2319.

40. J.M. Koene, G. Sundermann, and N.K. Michiels, "On the Function of Body Piercing During Copulation in Earthworms," *Invertebr Reprod Dev* 41 (2002): 35–40.

From the perspective of sexual-selection advocates, all these behaviors are *ipso facto* evidence of sexual conflict in hermaphrodites, conflict viewed as more severe than in separate-sex species. Yet, the emphasis on these phenomena is puzzling. Although few, if any, humans find these behaviors appealing, I have no idea what an earthworm's perspective is on body piercing. I could imagine a worm crawling through the dirt feeling quite relieved not be dragging genitals around and finding hypodermic insemination to be fast, efficient, and perhaps relatively painless. Moreover, it's not clear how general these behaviors are among the 65,000 species of hermaphrodites. Instead, researchers acknowledge copulation as a "genital 'handshake'"[41] and agree that "Reciprocity constitutes the prevalent mating mechanism among simultaneous hermaphrodites,"[42] suggesting that these behaviors are exceptional.

To sexual-selectionists, a hermaphrodite is the ultimate split personality, a body at war with itself, testes against ovary, left brain locked in combat with right brain. In a *Nature* report about penis fencing, Michiels and Newman state, "For males, injecting sperm offers direct access to eggs, whereas females bear the costs of wound healing and lose control over fertilization. Selection on females to avoid these costs must be strong but, in a hermaphroditic population, individuals can weigh the costs of being stabbed against the benefits of stabbing others."[43]

Interviews from a news article "Hermaphrodites duel for manhood" in *Science News Online*[44] reveal the thinking behind the *Nature* report by Michiels and Newman: "Talk about a battle of the sexes. Researchers have found hermaphroditic flatworms that rear up, expose their stubby penises, and literally duel. 'The interesting thing,' says Michiels, 'is that hermaphroditic partners run into conflicts because they usually have

41. N.K. Michiels and L.J. Newman, "Sex and Violence in Hermaphrodites," *Nature* 391 (1998): 547.

42. Nils Anthes and Nico K. Michiels, "Precopulatory Stabbing, Hypodermic Injections and Unilateral Copulations in a Hermaphroditic Sea Slug," *Biology Letters* 3 (2007): 121–124.

43. Michiels, "Sex and Violence in Hermaphrodites," 547.

44. S. Milius, "Hermaphrodites Duel for Manhood," *Science News Online*, 14 February 1998, http://www.sciencenews.org/pages/sn_arc98/2_14_98/fob2.htm.

identical but incompatible interests.' People may not realize how simple many human sexual conflicts are in the grand scheme of nature." Yet Michiels acknowledges, although does not discuss, other behaviors. The news article continues, "The dueling worms illustrate one extreme of hermaphroditic difficulties—both partners vying for the male role—but other flatworms have the opposite problem. 'Individuals have to 'beg' to receive sperm,' Michiels says. In these species, hermaphrodites project what looks like a penis but reaches out to a partner's male organs and withdraws sperm." The news article continues, "Hermaphrodites, which share the same sexual interests and strategy, may be more likely to evolve physically damaging sex, Michiels speculates. For the marine flatworms, particularly aggressive duelers may produce more offspring than so-so stabbers. 'It results in a kind of escalation,' he says." The theme of escalation is not limited to interview comments. In their *Nature* report, Michiels and Newman state, "Hypodermic insemination, when present, allows hermaphrodites to skew sexual interactions in favour of sperm donation, fuelling an evolutionary arms race between strike and avoidance behaviour" and Anthes and Michiels also state, "our behavioural evidence is in line with the hypothesis that mating in *Siphopteron quadrispinosum* [a sea slug] represents conflicting rather than complementary mating interests between mates."[45]

These behavioral studies have been supplemented with mathematical modeling purporting to show that an enhanced evolutionary escalation to violence occurs in hermaphrodites compared with separate-sex species.[46] The authors write, "Hermaphrodites can outweigh losses in the female function by gaining paternity .... Hence, hermaphrodites are more prone to be caught in costly escalations than gonochorists." Wife beating in hermaphrodites is meaner than wife beating in separately sexed animals. But then the authors acknowledge they have considered only one side of the issue: "Only harm to the female function

45. Michiels, "Sex and Violence in Hermaphrodites," 547.
46. Nico K. Michiels and Joris M. Koene, "Sexual Selection Favors Harmful Mating in Hermaphrodites More Than in Gonochorists," *Integrative Comparative Biology* 45 (2005): 473–480.

was considered . . . harming the male function was not considered here, as it is a form of male-male competition, which cannot take place between a female and male sexual partner." Sexual conflict as used by these authors is code for males inflicting pain on the female without considering the reverse possibility. Overall, the narrative of enhanced sexual conflict in hermaphrodites relative to separate-sex species is a tall story supported with selected examples and superficial modeling.

More generally, the discourse presently taking place about universal and severe sexual conflict in hermaphrodites foreshadows the difficulty some will experience with the concept of cooperative teamwork during reproduction, as discussed beginning with the next chapter. Even the unitary body of a hermaphrodite has been dissected into male and female components at war. Imagine how difficult it will be later to grasp that a male and female robin tending a nest together, two individuals in distinct bodies, are working as a team to raise young, that they are not subtly at war with each other. Imagine how far-fetched the idea of cooperation must seem to those devising stories of sea-slug violence.

My lab has committed to work in the near future on two further topics pertaining to the genetic system for sexual reproduction. One is why gametes bother to remain as haploid single cells, egg and sperm, that fuse to reconstitute a multicellular diploid embryo that develops into an adult. Alternatively, the gametes could set off on their own for a while, grow into multicellular organisms themselves, and co-occupy the habitat with their multicellular diploid parents. This life cycle is called, "alternation of generations" because the descendants from an individual alternate between haploid and diploid multicellular phases.

Priya is presently working on this problem as her thesis draws to a close.[47] The idea being investigated is that an advantage of the haploid phase lying between two diploid phases is to provide "gamete amplification." The rationale for gamete amplification is that if a diploid cannot

---

47. Priya Iyer and Joan Roughgarden, "Alternation of Haploid and Diploid Generations: Evolution by Gamete Amplification," *Evolutionary Ecology Research* (2009): in press.

make a sufficient number of gametes itself, it will make more in the long run by producing spores that set up house and make many more gametes that later fuse with one another.

Another topic my lab will be investigating pertains to the concept of what an individual is in evolutionary biology, with particular reference to the extended mycelium of fungi mentioned in Chapter 1. Recall the 2,200 year old gigantic mycelium that spans 2,400 acres site in eastern Oregon. This situation suggests an entirely different view of what an evolutionary unit is. Traditionally in evolutionary theory, the entity that evolves is a population, a population of individuals. But the mycelium example suggests that an evolutionary unit might instead be a gigantic individual who lives forever as a whole, but whose interior components come and go through time, so that the entire body is continually molding in response to surrounding conditions and is in turn affecting its surroundings. This truth-is-stranger-than-science-fiction view of an evolutionary unit would seem to be accurate for gigantic distributed organisms like fungi.

Last year, Henri Folse joined my lab after working at the Harvard Center for Risk Analysis where he developed Monte Carlo simulations of cervical and colorectal cancer to evaluate intervention strategies. Henri was an undergraduate at Harvard in applied mathematics and received a Masters degree in biology where he developed a spatially explicit simulation for a population of plants attacked by herbivorous insects. With this theoretical background, Henri decided to investigate how to develop the super-organism metaphor for an evolutionary unit in terms of a mathematical model, emphasizing how the dynamics of the nuclei within the mycelium are coupled to the growth and expansion of the super organism.

An important major question in Henri's research will be whether the nuclei within the mycelium separate themselves into individuals or pairs, with septa between them, or whether they exist together in a commons. The protoplasm within the mycelium can be viewed as a commons, like the open range of the American West, or can be fenced in, like individual ranches. In biology, one of the most basic of all axioms is the "cell theory"—that a multicellular organism consists of cells, each with a single nucleus. Why don't cells have more than one nucleus apiece? Well, in a fungal mycelium, two nuclei can shack up together or multiple nuclei may cohabit a region of protoplasm as a commune. Henri's research will initiate the theoretical study of issues pertaining to the origin of the cell theory of life and to the origin of biological individualism.

Overall, the social selection project springs from the philosophical premise that nothing in biology is necessary or "normal," and everything is contingent and evolutionarily derived from some primitive state, even though some derived conditions may be exceedingly common today. The task of science is to understand the circumstances in which each contingency evolves.

The Social System for Sex

SEVEN The Behavioral Tier

What about mammals, birds, lizards, frogs, fish, and all the bigger animals that surround us when we walk outdoors? Most examples mentioned during the previous part of the book, the part that concentrated on the origin of the genetic system for sex, involved organisms that are probably unfamiliar, organisms from single-celled algae to sea slugs. What about cooperation and sex among our furry, feathered, or scaly relatives? This part of the book is about the behavioral system for sex and is oriented to vertebrates, although not restricted to them. Snails and shrimp may well be described by the ideas we'll now take up, but still, in the mind's eye, the focus is on the behavior of the sort of animals most familiar to us since we were children.

I recall in 2004 wondering how possibly to go about devising a new theoretical approach to reproductive social behavior that would replace sexual-selection theory. Two ideas seemed foundational to a new theory of social selection, although initially I didn't know how to put them together.

The first foundational idea to social selection is the concept of "reproductive transactions," which I first encountered when reading the work of the behavioral ecologist, Sandy Vehrencamp, while doing research for *Evolution's Rainbow*. Her ideas led me to think of an animal reproductive social group as analogous to a firm in economics. The firm's product is offspring. The firm's employees might include relatives, as when a son inherits the company from his father, but might not, as when a company hires workers by placing an advertisement in the newspaper. Vehrencamp focused on insect colonies with multiple queens, a situation that is problematic for the explanation of insect social behavior according to the theory of kin selection originally developed by W. D. Hamilton, because that explanation envisions one queen as the parent of all the workers.

The basic idea of reproductive transactions is that an animal helps another in exchange for access to reproductive opportunity.[1] Some individuals, the employers, are envisioned to have control of reproductive opportunity and to pay out some of this opportunity to others, the helpers, who do not have similar access. In return for this paycheck, the helpers provide labor to assist the employers in their reproduction. The employer and helpers together constitute a reproductive social group, or team, analogous to a firm or household in economics.

Any inequality of reproductive opportunity initially available to different individuals is called a "distributional inequity" by economists. Distributional inequity may reflect territories that vary in exposure to predators or in the availability of food, water, and a mix of sunny and shady spots. Distributional inequity may develop because of inheritance, age, abilities, and luck.

1. S. Vehrencamp, "Optimal Degree of Skew in Cooperative Societies," *Amer Zool* 23 (1983): 327–335; H. Reeve, S.T. Emlen, and L. Keller, "Reproductive Sharing in Animal Societies: Reproductive Incentives or Incomplete Control by Dominant Breeders? *Behavioral Ecology* 9 (1998): 267–278; S. Emlen, "An Evolutionary Theory of the Family," *Proc Nat Acad Sci USA* 92 (1995): 8092–8099.

The sale of labor to produce offspring is especially profitable between relatives, leading to the formation of an extended family wherein an individual who does the breeding is a parent of helpers who remain at the nest. The value to a helper of assisting their parent's reproduction depends on their genetic relationship to the parent's offspring. The highest value accrues to offspring produced by the parent who are full brothers or sisters to the helper—in this case a helper may not bother with reproducing at all but let the parent do all the work, as in Hamilton's original idea of kin selection applied to social insects.[2]

An exchange of help for reproductive opportunity is possible even in the absence of a genetic relationship if the amount of reproductive access the helper is paid exceeds the reproductive opportunity the helper would have in the absence of supplying any labor. In retrospect, I think the lasting value of Hamilton's concept of kin selection does not much concern genetic relationship as a prerequisite for altruism or cooperation, but instead is the first evolutionary theory for transactions of reproductive opportunity. Genetic relationship affects the price of a reproductive transaction, whether a particular exchange profitable in terms of fitness to both parties, and therefore whether it takes place. The reproductive opportunity paid by an employer to a worker is called a "staying incentive," because this payment leads the worker to stay at the nest as a helper instead of leaving to start a new nest. In terminology to be used later, a staying incentive is a special case of a "side-payment" made between members of a team to keep the team intact and working toward a common goal.

Extended families form depending on supply and demand within the labor market. If demand for labor is tight—no jobs for youngsters outside of home, then even a small staying incentive will induce them to stay at home and join an extended family. With lots of opportunity outside of home, even a large staying incentive will not dissuade the youngsters from striking out on their own.

---

2. W. D. Hamilton, "The Genetical Theory of Social Behavior. I. II," *J Theor Biol* 7 (1964): 1–52.

According to this thinking, family structure is fluid, changing when opportunities outside the nest vary, and when the breeder changes mates which devalues the genetic paycheck a helper receives for their labor. Great distributional inequity causes reproduction to concentrate in a few individuals by mutual agreement between breeders and helpers, resulting in what is called a high "reproductive skew" that may amplify the initial inequity. If resources are evenly distributed, almost everyone breeds for themselves, and the social system has low reproductive skew. Sandra Vehrencamp termed these extremes as *despotic* versus *egalitarian* societies.[3]

When I first encountered Sandy Vehrencamp's ideas about reproductive transactions, I found them inherently appealing for their novelty and appreciated their usefulness in understanding social-insect societies with multiple queens. But I was especially interested because I was then grappling with how to organize the various forms of family and reproductive social groups I was trying to synthesize in *Evolution's Rainbow*. In species of birds and fish with two types of males, one is frequently a large territory holder and the other a smaller form that darts in to fertilize eggs laid in the large male's territory. Sexual-selection advocates call the small competitive male a "sneaker," often to the accompaniment of guffaws and snickers. This interaction is reasonably viewed as competitive, and susceptibility to incursions by the small males limits the territory size and power of the large male, in addition to any conflict between the large males themselves. This situation could legitimately be discussed as an extension of a sexual-selection narrative that preserves the ideology of male-male competition. Darwin's original account of sexual selection may not have envisioned more than one type of male, but two distinct competing types seems consistent with the sexual-selection picture of male-male dynamics.

On the other hand, a deep difficulty arises in interpreting bird and fish species with three types of males, one of which is actively solicited by the

3. S. Vehrencamp, "A Model for the Evolution of Despotic Versus Egalitarian Societies," *Anim Behav* 31 (1983): 667–682.

territory-holding male to join him in the dual courtship of females or in the dual guarding of eggs. The sexual-selection narrative must portray this third type of male, typically intermediate in size between the small competing male and the large territory holder, as somehow deceiving the large male into allowing it onto the territory whereupon it steals eggs that rightfully belong to the territory holder. I simply don't find the deceit narrative plausible because the assumption of deceit is always just stipulated and never independently demonstrated, and because the territory-holding male probably knows very well who is on his territory. Deceit narratives require great asymmetry in ability—the gullible victim and clever swindler, whereas natural selection should act to reduce such disparities if they ever exist. Instead of deception, the reproductive-transaction perspective would suggest that the courtship between the territory-holding male and the medium-sized male can be seen as a job interview rather than an act of deception. The idea of reproductive transactions invites thinking of the various types of males within the social group as a work team whose terms of employment need negotiation. Yet, the reproductive-transaction perspective still allows the possibility that other types of males are not participants in such reproductive teams and seek to confront it. Thus, the reproductive-transactions perspective adds to the interpretative possibilities—it does not deny competition, but does not require that all actions be twisted somehow into a form of competition.

Originally, this topic of how to think about multiple types of males was important to me when writing *Evolution's Rainbow*, because it allowed me to use a common framework to organize sex-role diversity into chapters about two-gendered families for reproductive groups containing one type of male and female and about multi-gendered families for reproductive groups in species with multiple types of males and/or females. But today, the greater importance of the reproductive-transaction perspective lies in how it approaches male-female relations. The basic male-female relationship can be viewed as a team. A male and female who are about to form a reproductive team have every conceivable incentive to work together. Therefore, sexual conflict, when it sometimes occurs, represents the failure to develop a working male-female contract—sexual conflict is not a fundamental starting point for under-standing male-female relations, but a secondary ending point represent-ing the contingent failure to negotiate a working agreement.

The reproductive-transaction perspective was quickly criticized by sexual-selection advocates.[4] Tim Clutton-Brock and colleagues argued that the employer may find the price of the staying incentives too high and not agree to pay. The helper should then abandon the nest and set forth alone. But the employer might try to coerce the helper to stay any-way. However, the employer might not be able to completely control the helper. The helper, therefore, breeds surreptitiously to whatever extent possible. Conversely, the net effect of the "helper" may actually hurt rather than help the employer. Does hanging around the nest and bring-ing in some food now and then yield a net benefit to the employer? In naked mole-rats, the breeding female aggressively activates "lazy" work-ers, suggesting a tension between employer and employee.[5] Either way, if the employer tries to coerce the worker or the worker exploits the em-ployer, the result is not a peaceful home, but a family at continual war.

Another of Clutton-Brock's objections is that mutual consent is be-yond what animals are capable of doing. Perhaps animals can't really ne-gotiate labor contracts among one another when people can't even do this very easily. And if they could negotiate labor contracts, do animals have social institutions to ensure that the contracts are honored? Do ani-mals have police to catch cheaters and lawyers to sue for breach of con-tract? These objections presume a human implementation of coopera-tion, and animals may have other, self-reinforcing mechanisms to realize cooperation. Indeed, I will argue later that pleasure, often physical, in the friendships that underlie cooperation has evolved to provide a self-reinforcing mechanism to achieve teamwork without requiring animal societies to contain counterparts of human social institutions.

4. T. H. Clutton-Brock, "Reproductive Skew, Concessions and Limited Control," *TREE* 13 (1998): 288–292; T. H. Clutton-Brock et al., "Cooperation, Control, and Concession in Meerkat Groups," *Science* 291 (2001): 478–481.

5. L. Keller and H. K. Reeve, "Partitioning of Reproduction in Animal Societies," *TREE* 9 (1994): 98–102.

I've suggested in *Evolution's Rainbow* that the objections sexual-selection advocates raised when the reproductive-transaction perspective was first proposed are the growing pains in the early stages of a new theory.

The second foundational idea to social selection involved the need to place genes at arm's length from behavior—to develop what I now term a *two-tier* approach to the evolution and development of social behavior. It's tempting to imagine that behavior has a genetic component similar to morphology. If there's a gene for eye color, or hair color, why not a gene for territoriality, for chirping sweetly in the morning, or for squawking fiercely during disputes? A gene might conceivably code for every type, or even every instance, of behavior. However, an animal's behavior obviously develops during its life reflecting experience in local situations. Some genetic component exists for behavior of course, but genes are remote from the minute by minute actions of animals. So, can we explain the evolution and development of behavior if not by recourse to genes?

This problem has been faced before. In the 1960s and early 1970s, a mathematical theory was developed by Robert MacArthur, Eric Pianka, Thomas Schoener, and Ronald Pulliam to predict what food an animal eats.[6] This theory was called the "optimal foraging theory." If you watch a bird, fish, or lizard, they obviously do not eat everything that crosses their field of view. They are choosey. So, how do they choose? Optimal foraging theory provides a mathematical modeling approach to animal decision-making in reference to food. An animal can decide what to chase down and catch, and what to ignore, according to the objective of maximizing the net energy it acquires throughout the period devoted to feeding. Feeding decisions are made by an individual animal, whereas in social behavior, as we'll discuss in more detail shortly, animals interact during their decision-making. Still, even though individual decision-making is simpler than social decision-making, the problem of keeping genes at arms length is present for both types of decision-making.

6. R.H. MacArthur and E.R. Pianka, "On the Optimal Use of a Patchy Environment," *American Naturalist* 100 (1966): 603–609; T.W. Schoener, "Theory of Feeding Strategies," *Annual Review Ecology Systematics*," 2 (1971): 369–404; H.R. Pulliam, "On the Theory of Optimal Diets," *American Naturalist* 108 (1974): 59–74.

Foraging theory was originally criticized for demanding too much of genetics. Suppose an optimal foraging model predicted that it was energetically optimal for some bird to chase down nearby bugs out to 1 meter away and ignore bugs beyond 1 meter, because they are not worth the bother. Well, if a gene exists that happens to code for chasing down bugs out to 1 meter while ignoring everything beyond, then such a gene would evolve in this situation over genes coding for any other chasing cutoff distance. But does such a gene exist? Surely not. And if the bugs became more plentiful after a rainfall so that chasing bugs out even to 1 meter was no longer worth the bother energetically, then would a different gene be needed for these conditions? In fact, birds, lizards, and fish do vary their decision-making day by day in accord with their experience. If bugs are hard to come by, then they chase bugs farther away than when bugs are plentiful. And their genes don't change day by day. The animals quickly learn somehow what the best use of their time is. So, foraging theory was then augmented with little behavioral rules that animals could plausibly follow, called "rules of thumb." These rules describe the foraging behavior minute by minute during the day.[7] By following the rule of thumb, a lizard's foraging behavior converges within minutes or hours, depending on the size and abundance of the prey in the environment, to the behavior predicted by optimal foraging theory, a theory that heretofore required assuming that optimal foraging was somehow encoded in the genes. According to these rules, behavior doesn't directly evolve, behavior *develops*. Although behavior changes developmentally minute by minute, the behavioral rule of thumb itself *evolves* from generation to generation, winding up with the rule that yields the best outcomes summed up day by day over the various circumstances that animals face. Hence, we have the basic idea behind a "two-tier" approach to the evolution of behavior. Behavior develops according to decision rules in a fast minute-by-minute time scale, whereas

7. Cf. J. Roughgarden, "The Sentient Forager," in *Anolis Lizards of the Caribbean: Ecology, Evolution, and Plate Tectonics* (Oxford: Oxford University Press, 1995); S. Shafir and J. Roughgarden, "Testing Predictions of Foraging Theory for a Sit-and-Wait Forager, *Anolis gingivinus*," *Behavioral Ecology* 9 (1998): 74–84.

the rules that underlie the decision-making evolve in a slow generation-by-generation time scale. Thus, the second foundational idea for social selection is to extend the two-tier methodology pioneered in individual decision-making to social decision-making.

Our two foundational requirements, therefore, stipulate that we must develop a theory of social selection that incorporates the possibility of reproductive transactions and that is structured in two tiers, developmental and evolutionary. In the balance of this chapter, we concentrate on the behavioral tier for processes that occur on a within-generation developmental time scale and leave the between-generation evolutionary tier to the next chapter. So let's get to work.

The first issue for behavioral-scale modeling is to decide on a useful state variable, that is, to decide on what quantity to follow through time that describes the state of the behavioral system. I chose the "time allocation" to various actions as the basic state variable.

When I was a graduate student in the early 1970s, I accompanied the herpetologist, Ernest Williams, in field work in the Dominican Republic. The purpose of the field work was primarily to collect lizard specimens to be brought back to the Museum of Comparative Zoology at Harvard. This work allowed time to pursue side projects, and I recall spending days documenting what is called the time budget of lizards. I would simply sit in the woods for several hours watching lizards and document what they were doing. The lizards from the species that live mostly midway up on tree trunks would move every 5 minutes or so, and would chase insects, display to one another, move to different positions, hide if a bird flew overhead, and so forth. The lizards from the species that lives near the base of tree trunks moved every 20 minutes or so—they are relatively immobile compared to the species that perches above them on the tree trunks. Males, females, big lizards, little lizards, lizards in the sun, and lizards in the shade, all carried out these types of actions with different frequencies. From this experience, I felt that the fraction of time per hour that is allocated to each type of action is directly observable and is characteristic of the local circumstances in which a lizard finds itself.

Because the time budget is readily observable, I decided to take it as the basic state variable to describe a social system. The task of the theory would then be to predict how the time allocation changes throughout the day and in different social and ecological situations. That is, if a lizard wakes up in the morning, for the first hour it may allocate 50% of its time during that hour to signalling with its neighbors and 50% of its time basking to heat up enough to chase insects, then in the second hour of the day, the lizard may shift its allocation to 25% signalling to neighbors, 25% chasing insects, and 50% to continually shifting position to gaze at different spots on the forest floor searching for bugs. And if a neighboring lizard catches a bug in its territory, the allocation might shift to 80% signalling back that the intrusion wasn't appreciated, leaving 20% to search for bugs. And so forth. The waxing and waning of the time-budget allocation to various actions, especially social interactions, for all the lizards in the area collectively describes the social system of the lizards in that area.

Mathematically, the time budget for a lizard at any particular time is a vector of allocations. Say there are three types of actions being considered: signalling to neighbors, chasing bugs, and shifting position to search for bugs. Then a specific lizard, say Joe, at 10:00 AM in the morning might have an allocation of (0.25, 0.25, 0.50). Suppose there are three more lizards in the vicinity, say Sally, Bill, and Helen, who interact with one another. Each of these lizards has an allocation vector, too. If we stack all the allocation vectors on top of one another, we form a table or matrix (spreadsheet) whose rows are the allocation from each individual to all types of actions and whose columns are the allocations by all individuals to each type of action. The entire matrix at any particular time then describes the state of the full social system at that time, and will be called the social-state matrix. The task of the behavioral-tier modeling is to predict how the social-state matrix changes throughout time. As the social system develops during the day, and from day to day, it may settle down to a steady state in which the allocations become a more or less unchanging set of time percentages. This might happen if the environment remains relatively consistent over several weeks and the animals have all signalled as much as they want to say to one another. This quieted-down situation will be called the equilibrium social state.

The selection of categories for classifying an animal's actions is a hypothesis itself, and must be conceived initially with some naturalist's intuition together with an eye toward utility. For example, splitting the category of "signalling to neighbors" into several would probably be useful, one for each pairwise interaction, say Joe with Sally, Joe with Bill, Joe with Helen, and so forth. Furthermore, splitting the category of "Joe signalling with Sally" into two would also be useful, one category for a signalling sequence initiated by Joe and one for a sequence initiated by Sally. These splittings would allow the social-state matrix to represent social teams, such as Joe with Sally and/or Joe with Bill, because the entries for time spent signalling between say, Joe and Sally would be greater than that for signalling between Joe and Helen, indicating that Joe and Sally could be a team together. On the other hand, the lizards I've worked with choose insects based more on size than on species. So, a butterfly the same size as a cricket should probably be lumped in one category. Although I'm disposed to be a lumper more than a splitter, I suppose a good procedure is to start with as many categories as is practical, to test statistically for differences, and then to lump those that are statistically indistinguishable. Moreover, the list need not be exhaustive to be useful. Special caution is needed because cultural, personal, ideological and theoretical biases easily enter. Such biases are then built in at the beginning of the project and remain indiscernible afterwards regardless of how rigorous subsequent procedures are.

Although the social-state matrix would be large in an empirical study, for modeling purposes it's best to begin with small matrices. For example, we've mostly concentrated on 2×2 matrices pertaining to, say, a male and female at a nest, together with two actions, say, incubating eggs and bringing food. The social-state matrix through time for this situation would describe how the male and female at the nest have developed a division of labor. For modeling, many extensions are possible as well, but small matrices are where to start.

After deciding on the social-state matrix as a useful state variable, I tried to dream up some equations to capture, I hoped, a picture of how animals might change each other's time allocation budgets through social interactions. But now the project stalled. The equations I dreamed up were, well, boring.

Then Meeko Oishi joined the project. I first met Meeko when she was a graduate student in the Department of Mechanical Engineering and her advisor, Claire Tomlin, was in the Department of Aeronautics and Astronautics in the School of Engineering at Stanford. Meeko took my upper-division undergraduate course in ecology. She took it for credit and did magnificently, and it was unusual to encounter a student from a department so distant from biology. But Meeko was truly interested in ecology and I got to know her a little bit. When Meeko reappeared in my office a year or so later, she had finished her Ph.D. degree and was now a postdoc with her former thesis advisor. Claire Tomlin had worked on issues, such as the design of air-traffic control systems to prevent aircraft collisions, as well as algorithms for cockpit automation—all topics as far removed from ecology as one could imagine, or so I thought. But Claire's lab was also involved in developing algorithms for decentralized optimization, which she describes as "meaning that several interconnected units act as local decision-makers and optimize local costs, coordinating with each other through constraints." Well, this is starting to sound a little bit like theory for the social behavior of machines, robots, piloted aircraft, or boats. So why not animals?

Meeko came to my office to say that she wanted to do a project in ecology and that Claire was interested in starting some collaborative work with biologists. Meeko suggested that the three of us get together to discuss possible topics. So, we had lunch together and I suggested a bunch of topics from molecular biology, physiology, and global change that seemed ripe for some theory of the type they had developed, as well as some persons to contact for possible collaboration. Then almost as an afterthought, I mentioned the work I had just started on the dynamics of social behavior wherein each animal was an agent and all somehow had to coordinate in pursuit of their welfare. I said I didn't think this would be a good area to start collaboration because it was controversial, the dialogue heated with lots of human social implication, and I had expressed definite views and would not be regarded as impartial. So, we left it at that. But several days later Meeko came by and said she had selected social behavior from menu of possible topics I had suggested. I was overjoyed, but didn't know how relevant the theory of decentralized optimization would prove to be. Since that time, Claire has gone on to develop projects with biologists on cell networks, and Meeko is now an assistant professor in the Department of Electrical and Computer Engineering at the University of British Columbia.

So, I showed Meeko the equations I had been working with and said they weren't going anywhere. Meeko entered them on her computer and agreed. Then as we were talking one day, she mentioned, "How about trying Nash bargaining?" What? My ears immediately perked up when I heard the name, John Nash, the same John Nash featured in the 2001 movie, *A Beautiful Mind*, starring Russell Crowe that was about the brilliant, but troubled mathematical economist who developed the basic theorems of game theory. The game theory I was familiar with was introduced by John Maynard Smith and summarized in his book, *Evolution and the Theory of Games*.[8] That book makes no mention of anything called Nash bargaining. So what does Maynard Smith present?

In the beginning of his book,[9] Maynard Smith sets the stage by explicitly defining behavior as a phenotype similar to a morphological phenotype. He writes, "A 'strategy' is a behavioral phenotype; *i.e.*, it is a specification of what an individual will do in any situation in which it may find itself." This decision enables his use of the language and approach one would adopt to determine whether a new gene, called a "mutant," will enter the gene pool, called "invading." He introduces a concept called the evolutionarily stable strategy (ESS) that indicates whether the evolutionary process has culminated in an endpoint. The ESS is an evolutionary counterpart of the equilibrium concept for competitive games introduced by John Nash in 1951 commonly referred to as the "Nash equilibrium."[10] Maynard Smith writes, "An ESS is a strategy such that, if all members of a population adopt it, then no mutant strategy could invade the population under the influence of natural selection." An ESS is not the gene itself; it's the strategy that the gene produces. A strategy is

8. John Maynard Smith, *Evolution and the Theory of Games* (Cambridge: Cambridge University Press, 1982).

9. Maynard Smith, *Evolution*, 10.

10. J. F. Nash, "Non-Cooperative Games," *Annals Mathematics* 54 (1951): 286–295.

an ESS if the gene coding for it is not displaceable by a gene for any other strategy drawn from a set of possible strategies. When this idea was proposed, it seemed eminently reasonable and has now been widely adopted in evolutionary biology and exported to other disciplines as well. However, the adequacy of the Maynard Smith approach needs to be reconsidered.

What are the strategies that Maynard Smith has in mind? His first example[11] is for the "Hawk-Dove Game." Maynard Smith writes, "Imagine that two animals are contesting a resource of value, V. By 'value' I mean that the Darwinian fitness of an individual obtaining the resource would be increased by V." He continues, "I suppose that individuals in a given context adopt one of two 'strategies'; for the time being, I assume that a particular individual always behaves in the same way. 'Hawk': escalate and continue until injured or until opponent retreats. 'Dove': display; retreat at once if opponent escalates. If two opponents both escalate, it is assumed that, sooner or later, one is injured and forced to retreat . . . . Injury reduces fitness by a cost, C." You get the flavor. Game theory is about fighting.

The other examples Maynard Smith introduces are similar in spirit, although different in detail. The "Battle of the Sexes" game that Maynard Smith credits to Richard Dawkins features two female strategies, "coy" and "fast" and two male strategies, "faithful" and "philanderer." The outcome is supposed to be cyclic. If females start out coy, then males should be faithful, but then females should become fast, which in turn causes males to philander, which finally causes females to return to being coy, starting the cycle all over again.

Still another example is the famous "Prisoner's Dilemma" game in which two prisoners are better off if both cooperate, each is best off and the other worst off if it defects while the other cooperates, and both do OK if both defect. The best and most desirable outcome is for both to cooperate, but the ESS is for both to defect. Because cooperation does not result, additional assumptions must be introduced to

11. Maynard Smith, *Evolution*, 11*ff*.

obtain a cooperative outcome such as repeated playing using a Tit for Tat protocol.[12]

Thus, what was in Maynard Smith's book seemed on the wrong track altogether for three reasons: it is exclusively single-tier in its conceptualization of behavior as genetic strategies, it assigns logical primacy to competition over cooperation instead of placing them on an equal footing to begin with, and it is bellicose in its choice of examples. Although the mathematics of game theory need not be presented using such loaded examples, collectively the examples confer a sense that game theory is all about conflict between selfish agents. As such, game theory didn't seem a promising theoretical framework to use when developing a new theory that would be focusing on reproductive transactions and teamwork, and that would be based in behavioral dynamics, not gene-pool dynamics. Wow, was I wrong.

Meeko's suggestion to try Nash bargaining led us to look further into John Nash's foundational work during the 1950 s on game theory, as well as later extensions—not on the competitive part but on the part pertaining to cooperation.[13] Maynard Smith, it now turns out, introduced only one half of game theory into biology, the half that pertains to competition. The other half of game theory pertains to cooperation, a half that I and most other biologists must blushingly admit to having been completely unaware of, even though this other half is relatively well known to game theorists in political science, economics, mathematics, and computer science. Meeko and I then began a crash tutorial for ourselves by reading and working

12. Robert Axelrod, *The Evolution of Cooperation* (New York: Basic Books, 1984).

13. John Nash, "The Bargaining Problem," *Econometrica* 18 (1950): 155–162; John Nash, "Two Person Cooperative Games," *Econometrica* 21 (1953): 128–140; Ariel Rubinstein, "Perfect Equilibrium in a Bargaining Model," *Econometrica* 50 (1982): 97–109; Ken Binmore, Ariel Rubinstein, and Asher Wolinsky, "The Nash Bargaining Solution in Economic Modelling," *Rand Journal Economics* 17 (1986): 176–188; Shinsuke Kambe, "Bargaining with Imperfect Commitment," *Games Economic Behavior* 28 (1999): 217–237; H. Peyton Young, "An Evolutionary Model of Bargaining," *Journal Economic Theory* 59 (1993): 145–168; H. Peyton Binmore, Larry Samuelson, and Peyton Young, "Equilibrium Selection in Bargaining Models," *Games Economic Behavior* 45 (2003): 296–328; S. Waslander, G. Inalhan, and C. Tomlin, "Decentralized Optimization via Nash Bargaining," in *Theory and Algorithms for Cooperative Systems*, Computers and Operations Research series, no. 4, eds. D. Grundel, R. Murphey and P. M. Pardalos (Singapore: World Scientific, 2004).

through examples drawn from a simple and elegantly written book on game theory[14] by the mathematician, Philip Straffin, which was originally recommended to me by a close personal relative who works in algorithmic game theory. What did we learn? What is cooperative game theory and how does it differ from competitive game theory?

The starting point for game theory, either competitive or cooperative, is the "payoff matrix," which is simply a table of what each player in the game earns depending on its action given what the others are doing. For example, let's take two birds at a nest, one male and one female. Say they have two possible actions, guarding the young from predators and collecting worms to feed the nestlings. The payoffs are in units of number of young successfully fledged times the probability of living through the season—this should approximately represent the increment to lifetime fitness accrued during a breeding season. If the nest fledges four nestlings and both parents are alive at the end of the breeding season, we assume that the payoff to each parent is four. But if one parent dies during the breeding season, say because of increased exposure to predation while foraging on the ground for worms, then let's assume that the number of young that can be credited to that parent is decreased to two.

All the possibilities are summarized in a payoff matrix shown in Table 3. Each entry in the table means (payoff to male, payoff to female). The top right (4, 2) means that when the male is guarding the nest and the female

Table 3   Hypothetical Payoff Matrix for Tasks at a Bird Nest

		Female	
		GUARD NEST	CATCH WORMS
Male	GUARD NEST	(1, 1)	(4, 2)
	CATCH WORMS	(2, 4)	(0, 0)

14. Philip Straffin, *Game Theory and Strategy* (Washington DC: Mathematical Association of America, 1993).

is catching worms, the male is credited with four young at the end of the season, and the female with two because she has a higher chance of dying from predation as a result of her foraging on the ground. Conversely, the lower left (2, 4) means that when the male is catching worms and the female is guarding the nest, he is credited with two young because of the hazard of predation while catching worms and she is credited with four young. The lower right (0, 0) means the nest is lost to predation from snakes and rodents if left unguarded, and the top left (1, 1) means that if both parents guard the nest and neither forages for food, then some nestlings starve while a few manage somehow to catch food on their own.

In a real-life application of this theory, one would want to fill in the entries with field data, and perhaps add more action categories, but please, don't make it more complicated than necessary. The simple table above is enough to illustrate how to think about this situation from a game-theory standpoint.

Now that we have a payoff matrix, it's time to play. But how do we play? What are the rules of play? In competitive game theory, the rule of play is for each player to act individually to maximize its own payoff, given what the other player is presently doing, and with the starting point and who goes first being set in advance (perhaps chosen at random with the flip of a coin). Suppose the game starts with both the male and female birds guarding the nest and earning payoffs according to the upper left of the payoff matrix. If the female moves first, she moves to the right because 2 > 1. Then the male has the next move, and he stays where he is because 4 > 0. In this case the female winds up catching worms and the male guards the nest. Alternatively, if the male has the first move, he moves down because 2 > 1. Then the female stays where she is because 4 > 0. In this case the male winds up catching worms and the female guards the nest. So, depending on who moves first, the game comes to an end at either the top right or bottom left corner, indicating that one bird is guarding the nest full time and the other is catching worms full time. Which is guarding and which is worm catching is arbitrary and depends on who got to move first.

Table 4   Movement Diagram for Individual Competitive Play at a Bird Nest

		Female		
		GUARD NEST		CATCH WORMS
Male	GUARD NEST		→	NCE
		↓		↑
	CATCH WORMS	NCE	←	

The same pair of outcomes happens if the game starts with both birds catching worms instead of both birds guarding the nest. And if the initial conditions are for one bird to guard and the other to catch worms, then they do not move from these positions. Thus the game always ends at either the top right or bottom left. These end-up points are called Nash competitive equilibria (NCE), or just Nash equilibria, for short. The competitive play of individuals in this way can be summarized as a "movement diagram" shown in Table 4. Notice that the arrows converge to one or the other of the two NCE's.

Playing a game thus represents how a social system *develops*. Equations describe how the initial social state quickly changes, minute by minute, into either of the two NCE social states assuming that each player continually adjusts its time allocation to increase its own payoff, given the present allocation of the other player. The NCE is the behavioral-time counterpart of Maynard Smith's ESS in evolutionary time, but doesn't require that the actions be genetic. The actions taken by the players develop as each player experiences the other's moves, and the culmination in an NCE makes no assumptions about whether "mutant" genes can "invade" the gene pool or not. Still, the philosophical spirit of individual competitive play leading to an NCE is identical to Maynard Smith's treatment. Moving to a two-tier formulation here is an improvement, because it eliminates the genetic determinism required in the standard single-tier formulations, but is philosophically unremarkable.

What is philosophically novel about behavioral-tier dynamics is that new ways of playing the game are opened up beyond those available with single-tier gene-pool dynamics. Think of the difference between playing chess and monopoly. In chess, the players can't talk or signal to each other and have to base their decisions about what move to make only on the present state of the chess board plus the record of previous moves that both players have made. In monopoly, the players can talk to each other and make deals. They can own Boardwalk as partners, and one can sell property to the other, and so forth. Such communication opens the door to cooperative game playing, not la-la let's-all-love-one-another cooperation, but cooperation that furthers self interest. This is the type of cooperation that the relatively unpublicized work of John Nash on cooperative game theory pertains to. Nash had labor-management negotiations in mind as the setup for his theory of "Nash bargaining" in cooperative game theory. Workers are not starry-eyed volunteers and management is not a sugar daddy; yet both need each other and must somehow come to terms. During the negotiations labor can threaten to strike and management to lock-out. The ideas Nash developed about bargaining during the 1950s are relevant to issues from labor-management negotiations, through playing monopoly, to who does what at a bird's nest.

Suppose then that our two birds have played competitively as individuals to an NCE state of the social system, say with the male guarding the nest and earning a fitness payoff for the season of four while the female is catching worms and earning a payoff of two. If the birds can communicate, as assumed possible in cooperative game theory, then what will they say to one another? The female should say that the current division of labor is unfair. To which the male replies, "tough."

But the female doesn't have to leave the matter there. She can force the male to bargain by establishing a credible threat. She can act against her own immediate interest and go on strike from further worm catching. Instead of flying off in the morning in search of worms, she can remain at the nest and contribute to guarding, even though this additional guarding is uselessly redundant. Well, the consequence of the female's remaining at the nest is that her own fitness earnings drop to one, as does the fitness earning of the male, according to the upper left entry of the payoff matrix. By remaining at the nest, she is acting against her own self interest and also hurts the male. Indeed, she can even play to hurt the male as much as possible. If she remains at the nest, the male should respond by foraging for worms himself. But to continue hurting him, she can then follow him by foraging, too. So, that now the nest is being left completely unguarded. Whenever the male plays a move she can follow and negate it. The result of the female playing to minimize the male's earnings even while he is playing to maximize his earnings is called the "threat point"—it's the best the male can do when the female is playing to hurt him the most.

After some number crunching, the threat point works out to be that the male is reduced to earning at the rate 1.6 fitness units per breeding season, which occurs when the male allocates $2/5$ of his time to guarding and independently the female allocates $4/5$ of her time to guarding. Conversely, the male can turn the tables and play to hurt the female as much as possible even as she is playing to maximize her earnings, which because the payoff matrix is symmetric for male and female, works out to be the same. But remember, establishing a threat point, making it credible, is expensive to the party doing the threatening—the threatening party's demonstrated willingness to suffer, to persevere during the strike, is what makes the threat credible. By now, both parties have an incentive to stop the fuss since each is receiving only 1.6 units of fitness per season which is worse for both than either NCE. But how should they stop the fuss? They shouldn't return to the original NCE, because the threats and counter threats will start all over again. Instead, they should somehow compromise. What then is the best compromise? Herein enters the concept of Nash bargaining, or more specifically, Nash arbitration—it offers a definition of what the optimal compromise is.

Nash proposed the concept of the Nash bargaining solution (NBS) based purely on abstract thinking, regardless of how the parties might go about arriving at the optimal bargain. That is, start by figuring out where the bargaining ought to wind up and later figure out a scheme that will lead there. So what should the ideal compromise be?

The first requirement is that all the win-win solutions should have been exhausted, leaving as candidate solutions only those which require hard choices. Also, a candidate solution must yield something better than what is being attained at the threat point. Otherwise why bother? These acceptable candidate solutions form what is called the "negotiation set."

Second, the optimal bargain should not depend on the units chosen to measure the payoffs, and we should be able to change the units by multiplying with a factor and/or adding a constant and still come out with the same answer as to how much time each bird should spend nest guarding and worm catching. This requirement allows the fitness payoffs to be specified as relative fitnesses.

Third, if we do happen to use units that render the problem symmetric, as our bird nest payoff matrix already is, then the optimal solution is a 50:50 compromise between them. A problem that's not symmetric can be transformed into one that is by changing units.

Fourth, irrelevant information shouldn't affect the solution either. This may seem obvious but has actually been the requirement that has attracted the most critical discussion over the years. Here's why the discussion arises. Suppose the birds can't actually spend 100% of their time guarding the eggs, say because they must forage for at least some worms to feed themselves. Only the excess of what each needs for his or her own maintenance is available for the nestlings. Well, should this new information change the optimal bargain between the birds because they can't actually carry out some of allocations we were originally assuming they could? It depends. If the threat point isn't changed by this new information, then no, the new information is irrelevant. However, if the threat point is changed by the new information, then yes, the optimal bargain does change. But if the new information does not translate into changing the threat point and is only manifested in making the new setup a subset of the original setup, then the optimal bargain doesn't change.

These four requirements are the four "axioms" of Nash arbitration and they uniquely specify what the optimal bargain is. For computation purposes, Nash showed that the optimal compromise is uniquely found by maximizing the *product* of the individual's net payoffs relative to their threat point. The product is maximized, and not the sum, because the solution that maximizes the product doesn't change when the payoff units are changed, whereas the solution that maximizes the

Table 5   Hypothetical Asymmetrical Payoff Matrix for Tasks at a Bird Nest

		Female	
		GUARD NEST	CATCH WORMS
Male	GUARD NEST	(0, 1)	(4, 2)
	CATCH WORMS	(2, 4)	(0, 0)

sum does change when the payoff units are changed. Therefore, maximizing the sum won't satisfy requirement two, whereas maximizing the product will.

Because our bird nest payoff matrix is symmetric to begin with, we know that the optimal compromise will be a 50:50 division of labor. Suppose we change the original payoff matrix so that it is not symmetric. For example, a male whose body size is larger than the female's automatically requires more food for metabolic maintenance than a female. The top left entry of the payoff matrix could be changed to indicate this supposition. Instead of (1, 1) for the payoff to male and female when both remain at the nest, we could take (0, 1) to indicate that the male starves if he does not leave the nest and search for worms, whereas the female can sustain herself with the occasional small bug she can catch while sitting near the nest. The new payoff matrix is then shown in Table 5.

To find the division of labor that represents the NBS for this case, we compute the new threat points, which takes some number crunching. The threat point for the male, which is the maximum the male can earn when the female is trying to hurt him the most, is now $1\frac{1}{3}$ fitness units, which is attained when he is guarding $\frac{1}{3}$ of the time and independently she is guarding $\frac{2}{3}$ of the time. If he tries to turn the tables, the threat point for the female, which is the maximum the female can earn when the male is trying to hurt her the most, is now $1\frac{3}{5}$, which is attained when the male is guarding $\frac{4}{5}$ of the time and independently the female is guarding $\frac{2}{5}$ of the time. What's interesting is that the female can now threaten the male more than the male can threaten the female, because the female can drop the male's earning down to $1\frac{1}{3}$, whereas the male can lower the female's earning down to only $1\frac{3}{5}$. Therefore, the female can drive a harder bargain than the male.

To find the optimal bargain we have to set up the formula for the product of the male and female earnings relative to their threat points. Specifically, for the male, we take the difference between the earnings that come from his choice of time allocations and his earnings at the threat point, and also for the female, we take the difference between the earnings that come from her choice of time allocations and her earnings at the threat point. Then we multiply these differences. Finally, we determine the set of male and female time allocations that maximizes this product. Although this computation would be tedious by hand, it's instantaneous on a computer. The result is that at the optimal bargain, the male earns 2.867 units of fitness per season and the female earns 3.133, and these earnings are attained when they *jointly* carry out the following allocations: 43% of the time the male is nest guarding while the female is worm catching, and 57% of the time the male is worm catching while the female is nest guarding. The female gets a better deal than the male because she has a stronger bargaining position—she gets to safely nest guard more than the male, and the male must dangerously worm catch more than the female, all because the female can drive a harder bargain. These sex roles—the female mostly tending the nest and the male mostly seeking food to bring back to the nest, emerge from the comparative bargaining positions of the male and female in these ecological circumstances. The roles would be reversed in different circumstances if the relative bargaining position of male and female were reversed according to the payoff matrix in those circumstances.

Readers must surely be forgiven for thinking it is far beyond the meager capacity of bird brains to compute an optimal compromise about how to spend their time. But we've been here before. Recall the rule-of-thumb discussion about how to forage optimally. The mathematical formula for the optimal foraging cutoff distance is a complicated expression involving a cube root. Do you think a lizard can compute a cube root in its head? I can guess your answer. But by following a simple stop-and-go rule as each potential prey item appears, a rule whose outcome changes as the lizard acquires foraging experience throughout the day, the lizard's behavior quickly converges on the optimal solution without having to bother calculating cube roots. If I were in the woods eating bugs, too, I wouldn't need a formula with cube roots as a crutch to understand what's going on. I could learn how to forage for myself. But I do use the cube-root formula, because I don't personally fancy a diet of ants, caterpillars, and crickets, and doubt that many readers do either. So, are there any simple rules or procedures to follow that will lead two birds at a nest to attain the optimal compromise about how to allocate their time to nest guarding and worm catching? Yes, there are three procedures (and probably more) that the birds can follow and wind up with an optimal compromise.

One procedure is called the "war of attrition." This procedure retains the spirit of individual competitive play and yet permits a cooperative outcome. A good illustration involves the division of space into territories.[15] Imagine two male lizards perching on adjacent trees. The problem is to arrive at a location for the territorial boundary between them. The value of the territorial space is highest for spots near where a lizard regularly perches and drops off for spots increasingly far away. Both lizards are strongly motivated to defend the space nearest to them, but their willingness to persist in territorial defense declines with distance away from the perch position. If lizards could simply signal to one another how much they value every piece of space between them, they could agree to each's owning the space that was more valuable to itself than to the other. The boundary would then occur at the spot where pieces of space changed from being more important to one of the lizards and switched to being more important to the other lizard.

Such utopian decision-making is conceivable, I suppose, but when watching lizards, one frequently sees them chase each other back and forth whenever one trespasses on the other's territory. This observation

15. Henrique Pereira, Aviv Bergman, and Joan Roughgarden, "Socially Stable Territories: The Negotiation of Space by Interacting Foragers," *American Naturalist* 161 (2003): 143–152; see also: Judy A. Stamps and V. V. Krishnan, "A Learning-based Model of Territory Establishment," *Quarterly Review Biology* 74 (1999): 291–318.

prompts the idea that a lizard reveals the value it places on a piece of space by demonstrating how long it will fight to defend the space. The lizard who gives up defending a piece of space loses it to one who is willing to continue fighting for it. By fighting over each piece of space between two lizards, the boundary can be found as the point between them where both fight equally, resulting in a draw. Exhaustion culminates in a cease-fire at a place that establishes the boundary between their territories. The static *outcome* is cooperative—the division of resources, but the dynamic *path* to this outcome is individualistic and competitive.

Another possible example of a war-of-attrition procedure to attain a cooperative outcome is the negotiation between the nitrogen-fixing bacteria, called rhizobia, which live in little bumps, called nodules, on the roots of legume plants such a peas. The bacteria take atmospheric nitrogen and bind it into amino acids which they send to the plant, and the plants fix atmospheric CO_2 into sugars, which they send to the bacteria. This biochemical symbiosis is hugely important to the world's agricultural production, as well as to the functioning of natural ecosystems. The basic question is how much energy the bacteria should commit to fixing nitrogen for export to the plants and how much energy the plants should commit to sugars for export to the bacteria.

Erol Akçay developed a physiological-time-scale model for the negotiation of this relationship.[16] As with behavioral-time-scale models, the physiological-time-scale model can later be embedded within an evolutionary model to form a two-tier theory for this symbiosis. Erol's idea for the physiology of the negotiation is that ordinary thermal fluctuations regularly take place in the amounts of material the plant sends to the bacteria and vice versa. These fluctuations can be thought of as physiological "offers." The plant or bacterium "rejects" an offer by shutting off its export in a "shock response" when the offer is unacceptable. An offer is unacceptable when the import one party is receiving drops below the point at which everything it makes is being allocated to its own growth with no excess available for export, otherwise an offer is accepted. By

---

16. Erol Akçay, and Joan Roughgarden, "Negotiation of Mutualism: Rhizobia and Legumes," *Proc Royal Society B* 274 (2007): 25–32.

alternating between accepting and rejecting random thermally-generated offers, the bacterium and plant home in on a compromise for the exchange of nitrogen and carbon between them. Erol showed that the outcome of this war-of-attrition process was a cooperative solution coinciding with the NBS that maximizes the product of the bacterial and plant growth rates. Again, the outcome is cooperative, but the path to the outcome is individualistic and competitive.

One might wonder though, whether cooperation is necessarily and always the incidental outcome of competition. Sexual-selection advocates would leave the matter here, content that cooperation has been subordinated to preeminent competition, leaving cooperation as a mirage shimmering above a desert of natural greed. Once again, the issue is not whether a preeminence of competition over cooperation is appealing or repugnant, but whether it is true. Do animals ever take cooperative paths to cooperative outcomes? Is complete and full cooperation a biological reality or are animals really stuck stumbling upon cooperative outcomes only after flopping down exhausted from the day's combat?

Glimmers of a new perspective appear in the casual and anecdotal observations we all have made of what can be described, I daresay, as "friendships" among our pets at home or anywhere we watch. You've surely seen mammals grooming each other or birds preening each other's feathers. Biologists dismiss this behavior as, "looking for bugs." How long does it take to find a few bugs? Perhaps mutual grooming is reciprocally pleasurable in itself with the activity prolonged to communicate a physical bonding. When you pet a cat or dog, aren't they enjoying your touch? I hope you aren't embarrassed to say that. When you hear birds talking, talking, incessantly, while they walk on the sand at the seashore or on mud flats at low tide, perhaps they're enjoying one another's cadence, marching in tune. When bats curl their tongues together, and female bonobo chimpanzees rub their groins together, they too are surely enjoying a physically intimate relationship.[17] And do

17. Cf. Frans B. M. de Waal, "Putting the Altruism Back into Altruism: The Evolution of Empathy," *Annu Rev Psychol* 59 (2008): 279–300.

animals really play?[18] How can animals take time out from the relentless striving to get ahead we're told that evolutionary success dictates? Just practice for later combat, working out in the boxing gym. Well maybe, but why not forming bonds of friendship that will underlie cooperation later in life? The ubiquity of physically intimate behavior in the animals all around us, behavior whose significance is always pooh-poohed by biologists, invites the heresy that friendships are real and important— important specifically as a mechanism to provide a cooperative path for attaining cooperative solutions to the dilemmas that animals face.

What then could possibly be the connection between physically intimate animal friendships and the mathematics of cooperative game theory? The clue lies in how to interpret the criterion Nash developed to compute his bargaining solution, the NBS. Although the existence of the NBS criterion, to maximize the product of the individual fitnesses relative to the threat points, has a perfectly fine mathematical justification based on the Nash axioms for what a fair bargain should be, there may be a more biological interpretation as well. We decided to name the product of the individual fitnesses as the "team fitness function."[19] The NBS is found by maximizing the team fitness function through joint coordinated action. So, Meeko, Erol (who had just joined the project), and I proposed equations for how the time allocations to the various actions would change through time if both animals were continuously working in coordination with each other to increase their team fitness. If the animals act jointly according to these new equations, instead of acting separately according to the older and standard equations, then mathematically,

18. Robert Fagen, *Animal Play Behavior* (New York: Oxford University Press, 1981); Robert Fagen and Johanna Fagen, "Juvenile Survival and Benefits of Play Behaviour in Brown Bears," *Ursus arctos. Evolutionary Ecology Research* 6 (2004): 89–102.

19. The team fitness function is not to be confused with the "team payoff function" or "team objective" as defined in the economic theory of teams. The team payoff function in the economics literature is the team-owner's payoff function or the coach's payoff function. In contrast, our "team fitness function" is a measure that combines the payoffs experienced by the team's players and not the payoff to their leaders or owners, because in our application, no one owns the team or is there a designated leader. Cf. for example, Theodore Groves, "Incentives in Teams," *Econometrica* 41 (1973): 617–631. I mention the economic approach to teams again in the next chapter.

Table 6   Movement Diagram for Cooperative Team Play at a Bird Nest

		Female	
		GUARD NEST	CATCH WORMS
Male	GUARD NEST	↘	↙
			NBS
	CATCH WORMS	↗	↖

their time allocations quickly approach the NBS instead of the nearest competitive equilibrium (NCE), as noted in Table 6.

We hypothesized that physically intimate contact produces two effects simultaneously: it keeps the animals coordinated and it conveys physical pleasure to each in the welfare of the other, supplying a personal physical motivation for joint welfare. Think of a basketball team. The athletes all hold hands, give high-fives before entering the floor, and stay "in touch" with eye contact. Then follows the joy of making a dunk with an alley-oop pass, the pleasure of a successful joint action, more fun than two foul shots which yield the same score. Team success involves a kind of synergy beyond the sum of utilities. When retired athletes who played a team sport are interviewed, they invariably recall missing the camaraderie the most. Successful teamwork involves *both* coordinated activity *and* the pursuit of a common team goal. We hypothesize that intimate physical contact stimulates physiological responses, hormonal and/or neural, that ensures coordinated activity and guides that activity to maximizing team fitness. Here, team fitness, as a product of the individual fitnesses, values the "synergy" to achieving a goal as a team beyond the sum of the values for the same achievement if attained individually. In 2006, Meeko, Erol, and I published these equations for attaining a NBS through team play in the journal, *Science*.[20]

20. Joan Roughgarden, Meeko Oishi, and Erol Akçay, "Reproductive Social Behavior: Cooperative Games to Replace Sexual Selection," *Science* 311 (2006): 965–969.

Team play based on physical intimacy provides an automatic check on whether both parties are in fact playing as a team. If one team mate deserts and tries to fly solo, the other team mate immediately senses the desertion and then reverts to the threat point. Team play is, thus, self-policing; there's no need to focus on detecting and punishing cheaters as there is in the conventional altruism theory, because there's no incentive to cheat as long as the threat points remain credible.

Is there any evidence that team play may actually occur? Some social psychologists have been speaking up to defend the importance of coordinated tactics during cooperation. In a large 2002 review article, Schuster[21] described lab studies in which *pairs* of rats were rewarded for coordinated shuttling within a shared chamber with unrestricted social interaction. Animals learned to work together with sensitivity to both task and type of partner. Schuster claims the significance of cooperative coordination has been "downplayed by learning theorists, evolutionary biologists, and game theorists in favor of an individual behavior → outcome perspective."

In the field, Judy Stamps and coworkers[22] published a study with blue-footed boobies in Mexico. Through exchanging a specialized communication signal (nestpointing) members of a pair communicate preferences, resolve disagreements, and ultimately decide jointly on where to place their nest. She also notes that "although many animal behaviourists are familiar with situations in which animal dyads appear to make joint decisions about space, little is known about the behavioural basis of this process." She continues with the conjecture that "perhaps one reason for the lack of interest in this topic is the lack of a framework for predicting the types of strategies and tactics that dyads might use" and she calls attention to some approaches developed for humans by social psychologists.[23]

21. R. Schuster, "Cooperative Coordination as a Social Behavior," *Human Nature* 13 (2002): 47–83.

22. J. Stamps et al., "Collaborative Tactics for Nestsite Selection by Pairs of Blue Footed Boobies," *Behavior* 139 (2002): 1383–1412.

23. D. Pruitt, *Negotiation Behavior* (New York: Academic Press, 1981); D. Pruit and P. Carnevale, *Negotiation in Social Conflict* (Pacific Grove: Brooks/Cole Pub, 1993).

Meeko worked out a particularly interesting example that was included in our *Science* paper. In the Eurasian oystercatcher, a species common on mudflats in which males and females are similar in appearance, some reproductive groups consist of threesomes with one male and two females, whereas most consist of one male and one female.[24] Moreover, the threesomes occur in two forms, aggressive and cooperative. In an aggressive threesome, each female defends her own nest, and the male defends a territory encompassing both females. The females lay eggs about two weeks apart. The females attack each other frequently throughout the day. The male contributes most of his parental care to the first-laid eggs, leaving the second nest often unguarded. In a cooperative threesome, the two females share one nest, both lay eggs in it together, about one day apart, and all three birds defend it together. The two cooperating females mate with each other frequently during the day, only slightly less often than they do with the male. The females also sit together and preen their feathers together.

Meeko developed the payoff matrix for this situation and showed that there were four Nash competitive equilibria, two of which correspond to both females attacking each other and the male helping one or the other of them, and two equilibria corresponding to one female, namely the one without the male's support, attacking the other female who does have his support, and the male helps one or the other. Meeko then determined the threat points and showed that the NBS was for both female birds to befriend each other and for the male to help each of them 50% of the time. This solution is attained through team-play dynamics, and represents the two females caring for their offspring jointly, and the male splitting his efforts equally between the offspring of both females, as automatically occurs when both females share the same nest. The physical

24. M. P. Harris, "Territory Limiting the Size of the Breeding Population of the Oystercatcher (*Haematopus Ostralegus*)—A Removal Experiment," *J Anim Ecol* 39 (1970): 707–713; B. J. Ens et al., *The Ostercatcher: From Individuals to Populations*, ed J. D. Goss-Custard (Oxford: University Press, 1996), 186–218; D. Heg and R. van Treuren, "Female-Female Cooperation in Polygynous Ostercatchers," *Nature* 391: (1998) 687–691.

intimacy between the cooperating females apparently underlies how the cooperative outcome has been attained.

As soon as our *Science* article appeared, sexual-selection advocates crashed down upon it like a ton of bricks, with 40 evolutionary biologists, mostly from the United Kingdom, combining to write 10 emotional letters of outrage in opposition.[25] The gang of 40 was attempting peer suppression, not peer review, bullying. In a follow-up, after conceding many of our objections to sexual selection, Clutton-Brock dismissed our paper's ideas by saying "Roughgarden's views are unusual."[26] The issue before us is not whether social selection is "unusual." Of course, social selection is unusual in evolutionary biology, because it emphasizes negotiation and offspring production rather than competition and mating. Instead, the issue before us is whether social selection is true and sexual selection false.

Our *Science* paper also briefly mentioned a third procedure for attaining the bargaining solution. A side payment, including reproductive transactions, can be represented as a modification to the payoff matrix that transforms the NBS into a competitive equilibrium, the NCE. If the advantaged party offers a side payment to the disadvantaged party so that the disadvantaged party receives the payoff it would have at the bargaining solution, then the incentive for disagreement dissolves. Returning to the payoff matrix in Table 3, suppose the bird guarding the nest who is receiving a payoff of 4 gives a unit of fitness to the bird who is catching worms and who had been receiving a payoff of 2. Then the payer's fitness drops to 3 and the payee's fitness rises to 3, which are the fitness values both would earn at the NBS.

The way a side payment is transacted depends on local details. The bird guarding the nest could do a bit of foraging around the nest itself,

25. Etta Kavanagh (ed.), "Debating Sexual Selection and Mating Strategies," *Science* 312 (2006): 689–697; our response appears on pp. 694–697.

26. Tim Clutton-Brock, "Sexual Selection in Males and Females," *Science* 318 (2007): 1882–1885.

Table 7   Hypothetical Payoff Matrix at a Bird Nest with Side Payments

		Female	
		GUARD NEST	CATCH WORMS
Male	GUARD NEST	(1, 1)	(3, 3)
	CATCH WORMS	(3, 3)	(0, 0)

supplementing the food brought by the long distance forager, thereby reducing the need for long trips to find food. The result would be Table 7 in which both NCE's return the same payoff as the NBS did. If the payer should cease making the side payments, the payee can revert to the threat point, and re-establish the incentive for bargaining.

I've wondered how a bird might sense how big a side payment to make. I suggest that to arrive at its appropriate side payment, a bird must feel a sense of pleasure in generosity and/or pleasure in seeing a just or fair division of resources—justice in the sense that resources are apportioned in accord with what each deserves as defined by the earnings each would realize at the NBS. I'm sure you've seen appeals from fundraisers for charities and religious organizations that assert it's emotionally healthy to be generous and that by being generous one attains a better state of mind, emotional satisfaction, and physical well-being. Indeed, neurobiological study has found that the sense of fairness fundamental to distributive justice "as suggested by moral sentimentalists, is rooted in emotional processing."[27] I suggest that birds, too, innately sense what side payments will contribute to their sense of pleasure and well-being.

27. Ming Hsu, Cédric Anen, and Steven R. Quartz, "The Right and the Good: Distributive Justice and Neural Encoding of Equity and Efficiency," *Science* 320 (2008): 1092–1095.

All in all, we have before us four methods of playing a social game which lead to the development of different social systems. Two methods involve cooperative paths to a cooperative outcome—team play and side payments. One method involves a competitive path to a cooperative outcome—the war of attrition. The fourth method involves a competitive path to a competitive outcome. How many social systems fall into these four categories is then an empirical question. An ideological commitment to the primacy of competition over cooperation should not preempt the empirical discovery of how common each of these types of social dynamics are.

I hypothesize that these four types of social dynamics arise as follows: the difference between using side payments rather than using team work to attain an NBS would seem to depend on whether the task being done by the social group is promoted or retarded by physical proximity. If the strength of two animals is necessary to subdue a prey or the eyes of two animals necessary to watch for predators from all angles, there's a functional advantage to physical proximity that would make teamwork the better mechanism for pursuing a cooperative path to a cooperative outcome. In contrast, if the task involves searching large areas for food or standing watch at different times, then side payments will allow the animals to carry out their tasks without the need for physical proximity and physical contact while the task was being carried out. In both of these cases, there is a team goal, but different paths to accomplish it.

In contrast, a competitive path to a cooperative outcome would seem to occur when there isn't any team goal, because the individual goals can't be bundled together into a common purpose, and also the individual goals are furthered when the individuals are physically separated. Finally, both the path and outcome are competitive if the organisms cannot discern a better outcome than the one they are experiencing at the NCE and/or they are unable to sustain the sacrifice needed to establish a credible threat point.

I hope future work will clarify how often cooperative or competitive outcomes are attained in social systems, and the paths by which they are attained. My hunch is that the commonness of social systems in which both the path and outcome are competitive has been greatly overrated and that the commonness of cooperative paths and outcomes has been greatly underrated. However, the evolutionary counterpart of the NCE, namely the ESS, does loom as the principal solution concept for evolutionary dynamics, and let's pursue this further in the next chapter devoted to the evolutionary tier.

EIGHT The Evolutionary Tier

The two-tier approach to the evolution of social behavior envisions that social interaction, including threat, negotiation, cooperation, and competition, takes place within a behavioral and/or physiological time scale with time measured in units, such as minutes, hours, and days, depending on the organisms' life span. Then the instantaneous fitness payoffs realized through these fast behavioral activities are cumulated over the life span to yield the generational fitness for use in population-genetic models defined on an evolutionary time scale. The evolutionary process itself is classical, even though the formulas for the generational fitness possess more detail than in conventional population genetics.

This two-tier approach applies to phenomena as diverse as the origin of multicellularity in which cells "agree" to team up as a common body, animals who agree to rear young together in family groups, and plants and microbes who participate together in symbioses. The equations for such varied phenomena differ in detail, because submodels for the behavior or physiology express the unique characteristics of each system being investigated. The game being played between two protozoa that might aggregate into a single body, between the rhizobia in a root nodule and its surrounding plant tissue, or between two birds tending a nest may be quite different, and the submodels for the payoffs accorded to these various forms of behavior and/or physiology respect the differences in their biology.

Here's an analogy for the overall setup. In a plaza near the Powell and Market Street cable-car station in San Francisco, one may see men playing board games such as chess. The plaza is the site of a population of games. Within each game, the position of each player ebbs and flows during the day as the game progresses. At the end of the day, each player takes their game board together with their winnings, goes home, and returns another day to play again with a new player. The "behavioral tier" describes the dynamics of the play at a game board by two (or more) parties, hour by hour within the day. The "evolutionary tier" describes the dynamics of the population of game-board players day by day throughout the year. The characteristics of the player-population evolve over the year as new player-types enter ("mutants" arise) and some old player-types disappear. A new "player-type" might possess a different payoff matrix than the existing payoff matrix, as though a player arrives at the plaza with a new and different game board, say a chess set with a different number of rooks, or a monopoly set with a few more Boardwalks. A new player-type might also possess a different rule book. If the players prosper from playing with a new payoff matrix and with new rules, then they slowly flood the plaza with a new game board or rule book, indicating that the game itself has evolved. And as the game evolves in this way, the minute-by-minute behavior changes, because the underlying payoff matrix that motivates the behavior is slowly changing through time.

After our *Science* paper that focussed on the behavioral tier appeared in 2006,[1] our lab turned to the evolutionary tier. In the spring of 2007, our lab group met each week and worked up a population-genetic model for how the payoff matrix could evolve through time. I presented the outlines of this work in the summer meeting of the Ecological Society of America (ESA) in San Jose, California. Then during the summer and fall of 2007, Erol Akçay and Jeremy Van Cleve, a graduate student in the same entering class as Erol who is studying in theoretical population-genetics, collaborated to develop an evolutionary model for the rules of play, specifically, whether to play cooperatively or competitively. I'll sketch the population-genetic model for the evolution of the payoff matrix now, and then introduce Erol and Jeremy's model for the evolution of the rules of play later in the chapter.

To readers unfamiliar with how population-genetic models are developed, the idea is simple. In principle, one constructs a spread sheet to total up the genes contributed to the next generation by each individual in the present generation. Then this totalling-up is repeated in another spread sheet for the next generation, and then the generation after that, and so on, thereby producing a projection of how the gene pool of a species changes through time.

In our case, we start with a payoff matrix, such as that between two birds tending a nest as illustrated in the previous chapter. Here we further envision that some genetic locus determines the properties of the payoff matrix. How would a gene determine the properties of a payoff matrix? By influencing the body size, metabolic rate, and/or foraging capabilities of the birds. A bigger bird might be able to lay a bigger clutch of eggs. But a bigger clutch would require more food, which means more time catching worms and being exposed to predation hazard. Hence, when compared with a smaller bird, the nest-guarding partner would have bigger payoff representing the larger number of eggs in the nest, whereas the worm-catching partner would have smaller payoff representing the greater exposure to hazard while foraging for more food.

1. Joan Roughgarden, Meeko Oishi, and Erol Akçay, "Reproductive Social Behavior: Cooperative Games to Replace Sexual Selection," *Science.* 311 (2006): 965–969.

Table 8   Hypothetical Payoff Matrices for Tasks at a Bird's Nest Where Both Male and Female Have Genotype A_1A_1, Producing a Small Body Size

		Female-A_1A_2	
		GUARD NEST	CATCH WORMS
Male-A_1A_1	GUARD NEST	(1, 1)	(4, 2)
	CATCH WORMS	(2, 4)	(0, 0)

Table 9   Hypothetical Payoff Matrices for Tasks at a Bird's Nest Where Both Male and Female Have Genotype A_2A_2, Producing a Large Body Size

		Female-A_2A_2	
		GUARD NEST	CATCH WORMS
Male-A_2A_2	GUARD NEST	(1, 1)	(5, 1.5)
	CATCH WORMS	(1.5, 5)	(0, 0)

Let's label the gene that produces the matrix in the previous chapter as A_1, and suppose the gene pool initially consists 100% of this allele, that is, the gene pool is initially "fixed" for A_1. Hence, both male and female have genotype A_1A_1 and their payoff matrix (taken from the previous chapter) appears in Table 8. Now we can envision a new allele, A_2, for a bigger bird that when present as a A_2A_2 genotype can lay more eggs, five instead of four. But an A_2A_2 bird must forage for a longer time than an A_1A_1 bird to feed these young and is more at risk from mortality, which reduces its expected fitness over the season to 1.5 from 2. If this gene were fixed in the gene pool, the payoff matrix could be that in Table 9.

Table 10   Hypothetical Payoff Matrices for Tasks at a Bird's Nest Where the
Male Has Genotype A_1A_1 and the Female Has Genotype A_1A_2,
and for Tasks Where the Male Has Genotype A_1A_2 and the Female
Has Genotype A_1A_1

		Female-A_1A_2	
		GUARD NEST	CATCH WORMS
Male-A_1A_1	GUARD NEST	(1, 1)	(4, 1.75)
	CATCH WORMS	(2, 4.5)	(0, 0)

		Female-A_1A_1	
		GUARD NEST	CATCH WORMS
Male-A_1A_2	GUARD NEST	(1, 1)	(4.5, 2)
	CATCH WORMS	(1.75, 4)	(0, 0)

Now, would this new allele, A_2, displace the original allele, should it
happen to arise as a mutation? How could we tell? To answer, we need to
spell out what all the possible payoff matrices are for all combinations of
player genotypes. For example, when the A_2 mutation first appears in
low frequency, it is present as a heterozygote together with the A_1 allele.
So, at this early stage soon after the A_2 allele has appeared, the payoff
matrices that are being played in the population are the original A_1A_1
male with A_1A_1 female according to Table 8, plus two new payoff matrices, a matrix for an A_1A_1 male playing with an A_1A_2 female, and a matrix
for a A_1A_2 male playing with an A_1A_1 female. Let's suppose the heterozygote has an medium body size resulting in payoffs that are between
those for A_1A_1 and A_2A_2 individuals, as illustrated in Table 10.

If we inspect these matrices, we can discern what will happen in the
gene pool after A_2 first appears. The baseline is set by what the A_1A_1
earnings are. The two rules of play are to play competitively as individuals or cooperatively as a team.

If playing competitively, the game ends at one of the Nash competitive equilibria (NCE) in the lower left or upper right corner of the payoff matrix. 50% of the time an A_1A_1 pair of genes finds itself in a female body and 50% of the time in a male body. Furthermore, as a male, 50% of the time the pair of genes winds up earning a fitness payoff of four if the bird containing the genes guards the nest, and 50% of the time the pair of genes winds up with a payoff of two if the bird containing the genes winds up being the one who catches worms. Averaging over these possibilities shows that the average payoff to an A_1A_1 male is three eggs successfully reared because it could be either the nest-guarder or worm-catcher. Similarly, the average payoff to an A_1A_1 female is three eggs successfully reared. Hence, an A_1 gene paired with another A_1 gene, once also averaged over the probability that it resides in a male or female body, experiences a payoff of three eggs successfully reared.

Alternatively, if playing cooperatively, the game ends at a Nash bargaining solution (NBS). The payoff experienced by an average A_1A_1 individual is again three eggs successfully reared because in this style of play male and female split the worm-catching and nest-guarding equally, again yielding a payoff of three to an average individual of this genotype.

Now let's compare the fitness earnings of an average A_1A_2 individual. Suppose they are playing competitively, with the game ending at one of the two NCEs. As either a male or female, the individual will earn on the average 3.125 eggs successfully reared.[2]

Alternatively, if they are playing cooperatively, then the payoff is again 3.125 eggs successfully reared. Thus, for either style of play, an average A_2 allele paired with an A_1 gene will wind up in a body and social circumstance that yields more offspring successfully reared than an average A_1 allele paired with another A_1, and therefore the A_2 allele will increase in the gene pool.

As the A_2 allele spreads, soon it will be found in bodies paired up with other A_2 alleles, and thereupon be homozygous. As this happens, even more payoff matrices are needed to specify all the additional genetic

2. $3.125 = (4.5 + 1.75)/2$

matchups that now occur. Although it's tedious to write out the complete set of all payoff matrices needed throughout the full evolutionary process, after an hour and a cup of strong coffee, one can list them all and then flesh out the complete spread sheet for how the evolution progresses. The evolutionary process will culminate with 100% A_2 indicating that A_2 has become fixed, and at this time the average payoff to an individual will have risen to 3.25.[3]

Conversely, after the gene pool has been fixed for A_2, should the A_1 rearise by mutation, it would not be able to increase, because an average individual carrying the now-rare A_1 allele will not yield as many eggs successfully reared as the baseline set by the now-fixed A_2 allele. Hence, the new payoff matrix in Table 9 is an evolutionarily stable payoff matrix, that is, an ESS payoff matrix, because an allele for another payoff matrix cannot increase when rare. In this sense, a two-tier approach for the evolution of social behavior retains the ESS concept, but locates the concept solely in the evolutionary tier where it belongs, whereas the two-tier approach employs a variety of stability concepts such as the NBS and NCE within the behavioral tier.

In this example, the social asymmetry increases through evolution, because a gene finds itself in both worm-catching and nest-guarding roles, and the gene confers more benefits when nest-guarding than it loses when worm-chasing, so it is beneficial overall, even though social asymmetry is increased. If the birds are playing competitively culminating at one of the NCEs, then within each instance of a game one of the birds suffers the whole cost of predation risk during its worm-catching while the other enjoys the whole benefit of a larger clutch of eggs. If the birds are playing cooperatively culminating at the NBS, then within each instance of a game both birds share the cost and benefits equally as both allocate 50% of their time to nest-guarding and 50% to worm-chasing.

After our lab explored how to frame the evolutionary tier within the language of population genetics focussing on the evolution of the payoff matrix, Erol and Jeremy went on to investigate the evolution of the rules

3. $3.250 = (1.5 + 5)/2$

of play, and specifically whether to play as individuals or teams, or somewhere in between. I've mentioned Erol's work previously and now is a good time to introduce him in more detail. Erol Akçay joined the lab in the fall of 2004 after obtaining his undergraduate degree from Middle East Technical University, Ankara, Turkey, with a double major in physics and biology, with honors. He arrived in time to witness the beginning of the social-selection project when I first presented the work with Meeko Oishi on Nash bargaining theory at the ESA 2004 meeting in Portland, Oregon. Erol then joined the project and became one of the coauthors when the work appeared later in *Science* in 2006. During his first year at Stanford, 2004 to 2005, Erol became very engaged with game theory, especially cooperative game theory, and his thesis, which was completed in June 2008, contains theoretical studies of how rhizobia and legumes negotiate their mutualism as already mentioned, how cooperative play evolves as will be mentioned shortly, how harmony and discord develop in animal family dynamics, and how extra-pair parentage forms and is maintained socially, as discussed in the next two chapters.

Jeremy Van Cleve also entered as a graduate student in the fall of 2004. He had obtained his undergraduate degree from Oberlin College with majors in mathematics and biology and Phi Beta Kappa honors. He joined the laboratory of Marc Feldman in theoretical population genetics and has published with Marc on the population-genetic consequences of genomic imprinting, which is a form of non-Mendelian inheritance. Jeremy also participated in my lab group meetings since the time he arrived and shared with Erol an interest in evolutionary game theory.

Readers will be observing that the cast of participants in the social-selection project has been growing. To keep everyone's contribution distinct, here is the chronology. The first participant was Meeko Oishi who introduced me to Nash bargaining, then Erol Akçay whose work has primarily been with cooperative game theory in the behavioral tier, then Priya Iyer whose effort has been on the evolution of the genetic system for sex, then the collaboration of Erol with Jeremy Van Cleve on the evolution of team play, and most recently, Henri Folse, who is working on the evolution of individuality with particular reference to fungi.

The lab-group discussions during spring 2007 were focussed on how the payoff matrix evolves, thereby changing the social behavior that develops from that. Whether to play competitively as individuals or cooperatively as teams was left as a stipulation, set in advance. The obvious next step, a fundamental step, is to see how a propensity for cooperative-team play evolves. Erol and Jeremy have developed what I think is a basic contribution to evolutionary biology in this regard. I suggested their paper might be entitled, "The Evolution of Love," but they wisely chose a more restrained title emphasizing the evolution of what they term *mutual regard*.[4]

Full team play as defined in our *Science* paper involves *both* pursuit of a team goal, which is the product of the fitnesses of the players, *and* coordinated action to attain that team goal. Erol and Jeremy decided to model an intermediate step, what I would call quasi-team play, in which the players act as uncoordinated individuals but nonetheless individually pursue goals that take the other's welfare into account. They introduce a two-tier formulation in which the individuals play to social equilibrium in the behavioral tier and then use the social-equilibrium payoffs in an evolutionary-tier model. These social-equilibrium payoffs depend on the extent to which each takes the other's welfare into account—they may benefit as individuals from a good social environment. Erol and Jeremy then compute the degree of caring for the other's welfare that is evolutionarily stable, an ESS. Erol and Jeremy's work demonstrates that cooperative-team work can evolve according to conventional population-genetic theory for evolution.

For the behavioral tier, Erol and Jeremy's introduce a formula for what might be considered as the personal motivations of the players. Each player's personal motivation to do some action depends on how much difference that action makes to their own personal objective. However, each's personal objective may take into account the other's welfare. So, an action by player A, say, that makes a big difference to player-A's

4. Erol Akçay et al., "On the Evolution of Mutual Regard," 2008: manuscript submitted for publication.

personal objective is quickly taken, even though player-A's objective contains both self interest as well as some concern for player-B's welfare. The formula Erol and Jeremy use is that each player's objective equals the payoff to itself resulting from the action, times the payoff to the other player raised to a power, β, where β is between 0 and 1. The value of β indicates the strength of the mutual regard. At one extreme, if β is 0, then a player's objective is purely selfish, interested in itself only, whereas if β is 1, a player's objective values the other's welfare as much as its own. Intermediate values of β indicate various balances between self interest and mutual regard.

Social interaction proceeds within the behavioral tier while both players carry out actions that further their personal motivations, allowing the possibility that those motivations may include concern for the other as represented by β. As both players continually act, the social system comes to a social equilibrium of actions. The genetic fitness being realized by each player at this social equilibrium depends on the parameter β. Next, β can evolve between generations. Erol and Jeremy showed that the ESS value of β is often greater than zero, and may equal 1, depending on the social returns that "investing" in mutual regard produces at the social equilibrium. Erol and Jeremy's model predicts the conditions in which the evolved mutual regard winds up between the extremes of purely selfish ($\beta = 0$) to completely empathetic ($\beta = 1$).

An important feature of Erol and Jeremy's model is that what evolves when $\beta > 0$ is genuine caring for the other, not make-believe empathy, not a clever deception camouflaging selfishness. Perhaps a new evolutionary psychology of the future might redirect its attention to the evolution of emotions in this way.[5]

The two-tier approach outlined previously also has been taken in some respects by other investigators; in the remainder of this chapter, I'd like to mention several related academic treatments.

5. Cf. Frans B.M. de Waal, "Putting the Altruism Back into Altruism: The Evolution of Empathy," *Annu Rev Psychol* 59 (2008): 279–300.

One formulation close to ours in spirit is the concept of an "inner game" nested within an "outer games" from Thomas Vincent.[6] This work originates in the discussions of coevolution between interacting species in an ecological community from the mid 1970s.[7] In Vincent's work, the inner game refers to models for the growth and decline of populations, such as the logistic and Lotka-Volterra equations in ecology textbooks, and the outer game refers to an evolutionary process by which the various coefficients in the population-change equations slowly evolve through time. Vincent's work clarified the way such two-level models should be formulated and led to improved models of species-coevolution.[8]

One difference between the inner-outer games formulation and the one proposed in this book involves the choice of whether to represent the outer evolutionary tier with explicit population-genetic equations or whether to represent the outer level as a game, in addition to representing the inner behavioral level as a game, too. The pros and cons to these approaches do not offer a clear choice. Representing the outer tier with explicit population-genetic formula requires simplistic genetic assumptions, such as "a gene for a payoff matrix," when obviously many genes would be involved. On the other hand, viewing the outer tier as a game requires faith that the actual genetical dynamics somehow work out to coincide with what a genetics-free model predicts. My own preference is for explicit genetics in the evolutionary tier that make the model assumptions transparent to everyone. Also, Vincent's research has not so far addressed problems where the inner tier is behavioral—he has focussed on ecological-community coevolution, not single-species social behavior.

6. Thomas L. Vincent and Joel S. Brown, "The Evolution of ESS Theory," *Ann Rev Ecol Syst* 19 (1988): 423–443; Thomas L. Vincent and Tania L. S. Vincent, "Evolution and Control System Design: The Evolutionary Game," *IEEE Control Systems Magazine* October (2000): 20–35.

7. J. Roughgarden, "Resource Partitioning Among Competing Species—A Coevolutionary Approach," *Theor Popul Biol* 9 (1976): 388–424; J. Roughgarden, "The Theory of Coevolution," in *Coevolution*, eds. D. J. Futuyma and M. Slatkin (Sunderland: Sinauer, 1983), 33–64.

8. Cf. J. Roughgarden, *"Anolis Lizards of the Caribbean: Ecology, Evolution, and Plate Tectonic,"* (Oxford: Oxford University Press, 1995), 103–120.

A second approach offering a two-tier flavor has been developed based on the genetics framework used for polygenic traits. Readers will be familiar with this framework, called "quantitative genetics," because it supplies the techniques to dissect how much a trait is brought about by genes, loosely speaking, and how much by "the environment". When you hear a geneticist say something like, "height is caused 50% by genes and 50% by environment" then you're hearing the outcome of an analysis using quantitative-genetic methods. The idea is to look at the distribution of, say, height among many people, and then partition the amount of variation in height among these people into that which can be predicted knowing their parent's height and the left over is assigned to the "environment," which would include, for example, nutrition. The "heritability" is the amount variation in height that is attributable to the height of the parents divided by the total amount of variation in height—this ratio lies between 0 and 1.

Alan Moore, Jason Wolf, Michael Wade, and their colleagues have extended this method to include social interactions in what they term the "interacting-phenotypes" approach.[9] That is, imagine that the some of the variation in say weight, is attributed to the parents' weight, some to nutrition, and some to whether your friends are big eaters. If you hang out with big eaters, you're likely to be heavy. By partitioning the total

9. David Queller, "A General Model for Kin Selection," *Evolution* 46 (1992): 376–380; Alan Moore, J. Edmund Brodie III, and Jason Wolf, "Interacting Phenotypes and the Evolutionary Process: I. Direct and Indirect Genetic Effects of Social Interactions," *Evolution* 51 (1997): 1352–1362; Jason Wolf et al., "Evolutionary Consequences of Indirect Genetic Effects," *Trends Ecology Evolution* 13 (1998): 64–69; Jason B. Wolf, Edmund D. Brodie III, and Allen J. Moore, "Interacting Phenotypes and the Evolutionary Process. II. Selection Resulting From Social Interactions," *Am Nat* 153 (1999): 254–266; Aneil F. Agrawal, Edmund D. Brodie III, and Michael J. Wade, "On Indirect Genetic Effects in Structured Populations," *American Naturalist* 158 (2001): 308–323; Allen Moore et al., "The Evolution of Interacting Phenotypes: Genetics and Evolution of Social Dominance," *American Naturalist* 160 (suppl): S186–S197 (2002); Stephen Shuster and Michael J. Wade, *Mating Systems and Strategies* (Princeton: Princeton University Press, 2003); Louise Mead and Stevan Arnold, "Quantitative Genetic Models of Sexual Selection," *Trends Ecology Evolution* 19 (2004): 264–271; Allen Moore and Tommaso Pizzari, "Quantitative Genetic Models of Sexual Conflict Based on Interacting Phenotypes," *American Naturalist* 165 (suppl): S88–S97 (2005).

variation in a trait into three components instead of two, that is, into the genetic contribution from parents, the contribution from those with whom one interacts socially, with the left over being assigned to the "environment," these researchers extended classic quantitative techniques to include social interactions.

As with other approaches, quantitative genetics has pros and cons. The pro is that it pertains to traits that are polygenic as would seem appropriate for most behavioral traits. The con is that the approach only works for short-term projections of evolutionary change, not for the full evolutionary trajectory starting with rare favorable mutations that end up being fixed.[10]

To see this, imagine a farmer selecting crops for a big tomato. At first, the selection progresses rapidly, because most of the size variation among tomatoes can be predicted by their parent plants. After a few years all the small size genes are weeded out, and the remaining variation is no longer predicted by the size of the parent plants, implying that the remaining variation must be classified as environmental, not genetic. During these years, the heritability will have changed from a nice high fraction, say $\frac{3}{4}$, and dropped to o by the end.

To use the quantitative-genetics approach, one multiplies the heritability times the strength of selection to obtain the projected outcome of evolution after one generation. The strength of selection measures the stringency of the "culling," which is how selective the group is that does the breeding—if nearly everyone breeds, the strength of selection is low, if only a particular few breed, the strength of selection is high. One can only predict the course of evolution one or two generations at a time for a given strength of selection because the ever-changing heritability must be continually re-estimated.

Similarly, the equations for the quantitative-genetic approach to social evolution assume that the matrix which is the multivariate interacting-phenotype counterpart to the heritability remains constant during the

10. Cf. J. Roughgarden, *Theory of Population Genetics and Evolutionary Ecology: An Introduction.* (New York: Macmillan, 1979; reprint, New York: Macmillan, 1987; reprint, Upper Saddle River, N.J.: Prentice Hall, 1996), 157–161.

projections of evolutionary change, and therefore can apply only for brief periods.[11] The authors are perfectly clear about this limitation, saying the assumptions are "made to permit us to investigate short-term evolution." Furthermore, the authors must also take the coefficient matrix describing the social interactions as constant also, together with assuming that "every individual engages in a single pairwise interaction." The approach is valuable nonetheless, and has led to artificial-selection studies on social status within populations of laboratory-reared cockroaches.[12]

Overall, the quantitative-genetic interacting-phenotype approach and the two-tier approach introduced in this book have different aims. The interacting-phenotypes approach aims at explaining relatively permanent traits, such as weight, body size, milk production, or an aggressive temperament. This book's two-tier approach instead aims to explain instances of behavioral actions, say an aggressive act to establish a threat point prior to negotiation, and to explain daily or hourly changes in the time allocated to different behavioral actions, such as the level of aggression (or cooperation) that occurs when a new individual enters the social group.

A major difference between this book's two-tier approach and the interacting phenotype approach, as well as some other approaches mentioned below, is that our two-tier approach is "bottom-up." We start with behavioral actions using game theory for a within-generation time scale and then build up to the between-generation time scale. The interacting-phenotype approach starts with the evolutionary time scale and unpacks the coefficients of a quantitative-genetic model down into pieces representing social interaction. If the aims are kept distinct, the two approaches should complement each other.

---

11. Cf. Alan, J. Moore, Edmund Brodie III, and Jason Wolf, "Interacting Phenotypes and the Evolutionary Process: I. Direct and Indirect Genetic Effects of Social Interactions," *Evolution* 51 (1997): 1352–1362.

12. Allen Moore et al., "The Evolution of Interacting Phenotypes: Genetics and Evolution of Social Dominance," *American Naturalist* 160 (suppl): S186–S197 (2002).

A third approach that shares a two-tier perspective is the theory of "biological markets," which was developed by Ronald Noë and Peter Hammerstein.[13] This approach focuses on the behavioral tier. Noë and Hammerstein write, "The exchange of commodities between individuals belonging to two different classes can be compared to the exchange of goods between two classes of traders in human markets."[14] Continuing, "As in human markets, the exchange rate of commodities on biological markets is determined by the law of supply and demand. In many cases a commodity in high demand will be exchanged for one in low demand." The dynamics of supply and demand for labor leads to "the division of labour in intra-specific cooperation, e.g. the relationship between the workload of helpers at the nest and the number of helpers relative to breeder pairs in a population." They interpret these social outcomes as "behavioural adjustments to local market situations" that "do not necessarily imply adaptations in an evolutionary sense." Noë and Hammerstein state further that, "we left cheating as a strategic option out of consideration, since we wanted to concentrate

---

13. R. Noë, "A Veto Game Played by Baboons: A Challenge to the Use of the Prisoner's Dilemma as a Paradigm for Reciprocity and Cooperation," *Anim Behav* 39 (1990): 78–90; R. Noë, "Alliance Formation Among Male Baboons: Shopping for Profitable Partners," in *Coalitions and Alliances in Humans and Other Animals*, Eds. A.H. Harcourt and F.B.M. de Waal (eds) (Oxford: Oxford University Press, 1992), 285–321; R. Noë, C.P. van Schaik, and J.A. van Hooff, "The Market Effect: An Explanation for pay-off Asymmetries Among Collaborating Animals," *Ethology* 87 (1991): 97–118; R. Noë and P. Hammerstein, "Biological Markets: Supply and Demand Determine the Effect of Partner Choice in Cooperation, Mutualism and Mating," *Behavioral Ecology Sociobiology* 35 (1994): 1–11; R. Noë and P. Hammerstein. Biological Markets," *Trends Ecology Evolution* 10 (1995): 336–339; R. Noë, "Biological Markets: Partner Choice as the Driving Force Behind the Evolution of Mutualisms," in *Economics in Nature: Social Dilemmas, Mate Choice and Biological Markets*, eds. R. Noë, J. A. van Hooff, and P. Hammerstein (New York: Cambridge University Press, 2001), 93–118.; P. Hammerstein, "Games and Markets: Economic Behavior in Humans and Other Animals," in *Economics in Nature: Social Dilemmas, Mate Choice and Biological Markets*, eds. R. Noë, J. A. van Hooff, and P. Hammerstein (New York: Cambridge University Press, 2001), 1–22; P. Hammerstein (ed.), *Genetic and Cultural Evolution of Cooperation* (Cambridge: MIT Press/Dahlem University Press, 2003); Peter Hammerstein, Edward H. Hagen, and Manfred D. Laubichler, "The Strategic View of Biological Agents," *Biological Theory* 1(2006): 191–194.

14. Noë and Hammerstein, 1994 op. cit., 1–3.

on the market effect itself. . . . To our minds the cheating option can safely be ignored in the large number of cases in which either the commodity cannot be withdrawn or changed in quality or quantity once it is offered on the market, or when cheating is effectively controlled."

Our two-tier approach in this book shares this de-emphasis on cheating and policing, as well as its emphasis on transactions. A difference is that our analysis uses cooperative game theory, including the Nash bargaining solution, as the key technical machinery for modeling, whereas Noë and Hammerstein rely on the Nash competitive equilibrium. As such, we emphasize friendship and mutual caring in our modeling of behavioral dynamics, whereas Noë and Hammerstein's interpretation is more suited to perfect-market dynamics. Finally, our modeling develops the evolutionary tier in more detail than Noë and Hammerstein have so far. I see our two-tier theory and the biological-markets theory as nicely complementary.

A fourth approach relevant to our two-tier theory pertains to our hypothesis of team-play dynamics within the behavioral tier. Philosophers have begun to discuss how to conceptualize team intention and team agency and how teams might be said to "reason."[15] This research can clarify and generalize what team play means. Our two-tier theory offers a biological context in which this research in philosophy might be applied. Our formulation is rather specific compared to the discussions taking place in philosophy, because we envision a specific team fitness function as a product of individual utilities, not a sum, and we envision that the pleasure of cooperation guides players to pursue a common goal with coordinated actions.

15. Michael Bratman, "Shared Intention," *Ethics* 104 (1993): 97–113; Robert Sugden, "Thinking as a Team: Toward an Explanation of Nonselfish Behavior," *Social Philosophy Policy* 10 (1993): 69–89; Robert Sugden, "Team Preferences," *Economics Philosophy* 16 (2000): 175–204; Robert. Sugden, "The Logic of Team Reasoning," *Philosophical Explorations* 6 (2003): 165–181; Natalie Gold, *Framing and Decision Making: A Reason-Based Approach* (D. Phil thesis, University of Oxford, 2005); Michael Bacharach, *Beyond Individual Choice: Teams and Frames in Game Theory* (Princeton: Princeton University Press, 2006); Natalie Gold and Robert Sugden, "Theories of Team Agency," in *Rationality and Commitment*, eds. Peter Fabienne and Hans Bernhard Schmid (Oxford: Oxford University Press, 2008), 280–312.

A fifth vein of literature consists of "the theory of teams" in economics and management science, a large literature that extends into the 1950s.[16] Much of this work pertains to the management of firms, and treats issues such as ensuring accurate and efficient information exchange between the employers and employees. The conceptualization of incentives and profit sharing in this literature may be helpful to developing a behavioral-tier theory for the "staying incentives" exchanged within an extended family or other reproductive social group between the member who controls resources and the others who act as pre-zygotic or post-zygotic helpers at the nest.

A sixth line of recent research similar to our two-tier theory also comes from economics where questions such as how consumer preferences develop and whether consumer preferences accurately reflect actual utilities have been studied.[17] This literature considers issues, such as the long-term benefit to optimism, which in the short term may represent an inaccurate assessment of the circumstances. This theory is similar mathematically to

16. J. Marschak, "Elements for a Theory of Teams," *Management Science* 1 (1955): 127–137; Theodore Groves and Roy Radner, "Allocation of Resources in a Team," *J Economic Theory*, 4 (1972): 415–441; Theodore Groves, "Incentives in Teams," *Econometrica* 41 (1973): 617–631; Theodore Groves and Martin Loeb, "Incentives in a Divisionalized Firm," *Management Science* 25 (1979): 221–230; Roger Myerson, "Incentive Compatibility and the Bargaining Problem," *Econometrica* 47: 61–74; K.J. Arrow and R. Radner, "Allocation of Resources in Large Teams," *Econometrica* 47 (1979): 361–385; Bengt Holmstrom, "Moral Hazard in Teams," *Bell J Economics* 13 (1982): 324–340; Susan I. Cohen and Martin Loeb, "The Groves Scheme, Profit Sharing and Moral Hazard," *Management Science* 30 (1984): 20–24; Martin L. Weitzman, "The Simple Macroeconomics of Profit Sharing," *American Economic Review* 75 (1985): 937–953; Felix R. Fitzroy and Kornelius Kraft, "Cooperation, Productivity, and Profit Sharing," *Quarterly J Economics* 102 (1987): 23–35; Kornelius Kraft, "The Incentive Effects of Dismissals, Efficiency Wages, Piece-rates and Profit-sharing," *Review Economics Statistics* 73 (1991): 451–459; Douglas L. Kruse, "Profit-sharing and Employment Variability: Microeconomic Evidence on the Weitzman Theory," *Industrial Labor Relations Review* 44 (1991): 437–453.

17. Larry Samuelson, "Introduction to the Evolution of Preferences," *J Econ Theory* 97 (2001): 225–230; Efe A. Ok and Fernando Vega-Redondo, "On the Evolution of Individualistic Preferences: An Incomplete Information Scenario," *J Econ Theory* 97 (2001): 231–254; Aviad Heifetz, Chris Shannon, and Yossi Spiegel, "What to Maximize if you Must," *J Econ Theory* 133 (2007): 31–57; Aviad Heifetz, Chris Shannon, and Yossi Spiegel, "The Dynamic Evolution of Preferences," *Econ Theory* 32 (2007): 251–286; Aviad Heifetz, Ella Segev, and Eric Talley, *Games Econ Behav* 58 (2007): 121–153; cf. also Werner Güth and Hartmut Kliemt, "Evolutionarily Stable Co-operative Commitments," *Theory Decision* 49: 197–221.

the formulation devised by Erol and Jeremy in their paper on the evolution of mutual regard.

A seventh study to mention appeared in 2007 by Lee Worden and Simon Levin[18] that begins with observations I certainly endorse. They write, "The widespread conflation of cooperation and altruism in the study of behavior reinforces a widely shared view of the world in which cooperation is a mysterious anomaly whose existence is difficult to explain because of the universal temptation to defect. Prisoner's dilemma and tragedy of the commons scenarios, which embody these assumptions, appear to the lay reader or student as authoritative scientific statements legitimizing a bleak Hobbesian picture of humans as selfish, greedy individualists whose antisocial tendencies must be kept in check by coercive social forces."

Worden and Levin then introduce a two-tier setup somewhat similar to ours in that they envision a population of games. In the prisoner's dilemma game, if both players are earning a certain payoff by cooperating (not tattling to the guards), then either can win an immediate gain by defecting (tattling to the guard), which causes the other to defect as well, (both are then tortured by the guards) leading to the lowest payoff overall. Figuring out how to avoid this rather depressing outcome has engaged game theorists for decades.

Worden and Levin allow the payoff matrix itself to evolve. They write, "The entries of the payoff matrix, which characterize each play in terms of how it interacts with all the other plays, are perturbed by small numbers to represent the introduction of a new, previously undiscovered play, which is a slight variant of one of the existing ones." They find that "the set of plays currently in use undergoes evolutionary change, and the payoff matrix undergoes gradual qualitative changes." The payoff matrices in their simulations wind up being ones in which mutual cooperation not only yields the highest payoff for both, as it did to begin with, but also the payoff for one or the other to defect are lower than

18. Lee Worden and Simon A. Levin, "Evolutionary Escape from the Prisoner's Dilemma," *J Theoretical Biology* 245 (2007): 411–422.

to continue cooperating. Thus, the incentive to defect disappears during the evolution, making mutual cooperation the stable outcome, stable even with competitive play—that is, after evolution mutual cooperation is a Nash competitive equilibrium in the new payoff matrix. Unlike our two-tier setup that relies on coordinated team play toward maximizing team fitness to yield a bargaining solution in behavioral time, Worden and Levin attain a cooperative outcome in evolutionary time solely through the evolutionary process.

An eighth approach that at first glance appears relevant to a two-tiered perspective has been well publicized by John McNamara, Alasdair Houston, and colleagues.[19] They offer a top-down approach, their "best-response" theory, which starts by modeling the evolutionary tier with game theory, and then unpacks the ESS solution into "negotiation rules" in the behavioral tier. The negotiation rules are supposed to culminate in the ESS. They write, "Most two-player games should be modelled as involving a series of interactions in which opponents negotiate the final outcome. Thus we should be concerned with evolutionarily stable negotiation rules rather than evolutionarily stable actions. The evolutionarily stable negotiation rule of each player is the best rule given the rule of its opponent." Continuing, "We present an analysis of negotiation in the context of a pair of animals feeding their young . . . a conflict of interest exists, with each parent preferring the other to work hard."

The emphasis on negotiation would lead one to suspect that there might be much in common between our two-tier approach and McNamara and Houston's. Alas, no. The negotiation rules they derive for the behavioral tier are necessarily individualistic and competitive, too. This exclusion of cooperative dynamics from the behavioral tier is deliberate and ideological and they argue that, as a matter of principle, competitive dynamics logically precede cooperative dynamics. They write, "We argue that noncooperative game theory provides an adequate basis for understanding sexual selection . . . . Bargaining does not require the assumption

19. John M. McNamara, Catherine E. Gasson, and Alasdair I Houston, "Incorporating Rules for Responding into Evolutionary Games," *Nature* 401 (1999): 368–371.

of cooperation and does not necessarily lead to it."[20] Thus, McNamara and Houston regard their emphasis on individualistic competitive behavior ("each parent preferring the other to work hard") as a positive feature, not a limitation.

The technical details in the best-response theory are troublesome too. Any existing behavioral configuration, such as how much effort the male and female are each providing to the young, is assumed be sufficiently close to an ESS solution so that the best response by one party to the state of the other party is a linear function of the difference between the current state and the ESS state. This linear approximation (a Taylor-series expansion) is unjustified and almost surely an inaccurate representation of actual give-and-take behavior, even if the animals are individualistic competitive fitness maximizers. Other aspects of their mathematical analysis have also been found flawed.[21] In my opinion, the best-response approach of McNamara and Houston is a nonstarter—an inadequate, static representation of the evolutionary tier and an ideological, inaccurate representation of the behavioral tier.

That said, the eight approaches cited previously, plus our own two-tier approach introduced in this book, support the hope that evolutionary theory is possible for social behavior in which the evolutionary tier is kept at arm's length from the behavioral tier. In biology there is dwindling support, if any, for stories of genetic determinism concerning behavior. Two-tier approaches collectively undercut proposals in behavioral ecology, evolutionary psychology, evolutionary anthropology, and evolutionary ethics that tie genes to specific instances of behavior.

Finally, I need to revisit the topics of group selection and kin selection to reiterate that our two-tier approach to the evolution of social behavior is fundamentally different from group selection. Readers familiar with how group selection was discredited during the 1970s and how individual-level selection was broadly endorsed may be surprised to learn how

20. John M. McNamara, Ken Binmore, and Alasdair I. Houston, "Cooperation Should not be Assumed," *Trends Ecology Evolution* 21 (2006): 476–478.

21. P. Taylor, and T. Day, "Stability in Negotiation Games and the Emergence of Cooperation," *Proc R. Soc London B* 271 (2004): 669–674.

much the scene has since changed. When I was a graduate student in the early 1970s, I recall being asked on a Ph.D. qualifier exam what I thought at the time was a leading question, which if I recall correctly, went something like this: "Which of the following is why group selection is wrong?" We were drilled to disavow group selection after the individual-selection proponent, G. C. Williams[22] critiqued the group-selection proponent, V. C. Wynne-Edwards.[23] The individual-selection premise holds that traits evolve because they benefit the individual who possesses them, whereas the group-selection premise holds that traits evolve because they benefit the group to which the individual belongs.

As of the early 1970s, group selection seemed empirically suspect—it just never happened, and was theoretically dubious—no models had been devised to show how it might work in terms of population genetics. Then models began to appear showing how group selection might result from differential survival between groups[24] and from differential contributions of offspring from groups to the population at large.[25] After the process became understood theoretically, group selection was soon demonstrated in the laboratory[26] and in the wild.[27] Nonetheless, I think it's fair to say that today the consensus still remains that group selection is rare relative to individual selection because the population structure has to satisfy restrictive conditions that seem rarely met. The matter appeared more or less settled until recently when the dispute between

22. G.C. Williams, 1966. *Adaptation and Natural Selection: A Critique of Some Current Evolutionary Thought* (Princeton: Princeton University Press, 1966).

23. V.C. Wynne-Edwards, *Animal Dispersion in Relation to Social Behavior* (Edinburgh: Oliver & Boyd, 1962).

24. I. Eshel, "On the Neighborhood Effect and the Evolution of Altruistic Traits," *Theor Popul Biol* 3 (1972): 258–277; J. Roughgarden, *Theory of Population Genetics and Evolutionary Ecology* (New York: Macmillan, 1979; Upper Saddle River, N.J.: Prentice-Hall, 1996), 283–292 (see illustration and numerical examples).

25. D.S. Wilson, "A Theory of Group Selection," *Proc Nat Acad Sci USA* 72 (1975): 143–146.

26. Michael J. Wade, "Group Selection Among Laboratory Populations of *Tribolium*," *Proc Nat Acad Sci USA* 73 (1976): 4604–4607.

27. R. Colwell, "Group Selection is Implicated is Implicated in the Evolution of Female-biased Sex Ratios," *Nature* 290: 401–403.

individual and group selection as major causes of evolutionary change has flared up again.

In the context of the individual versus group selection dispute, kin selection has always been taken as a form of individual selection—an animal forgoes reproduction itself to help a close relative reproduce and if the number work out right, the animal winds up leaving more copies of its genes in the next generation by this indirect route rather than by the direct route of producing the offspring itself. Altruism, that is helping another at a cost to one's self, is thus rendered as individually advantageous. This process must surely be the antithesis of group selection. Yet, the seeming clarity distinguishing kin selection from group selection has long seemed illusory. Numerous papers have pointed out how kin selection and group selection amount to the same thing, depending on how one looks at the matter.[28] Readers steeped in the individual-selection tradition might gasp, "How could this be?" To say that kin selection and group selection amount to the same thing, depending on one's point of view, is to say that black amounts to white, depending on one's choice of glasses. To say kin selection and group selection are the same seems mischievously to fudge an obvious distinction and to take the wind out of the sails of the ardent combatants on both sides.

So, here's why kin selection and group selection are intertwined to the point of being almost indistinguishable: if the members of a group stick together and reproduce through several generations, they automatically become more related to one another than they are to members of other groups. Suppose a group of animals originally banded together to catch food and watch for predators. Suppose the next group is miles away, so animals of the group tend to breed with one another. Then after a few generations, everyone in the group is related. At this time, one could say, well, the members are being altruistic in their joint foraging and patrolling for predators because they are helping one another's relatives, or

28. G. R. Price, "Selection and Covariance," *Nature* 277 (1970): 520–521; D. C. Queller, "Does Population Viscosity Promote Kin Selection?" *Trends Ecol Evol* 7 (1992): 322–324; P. D. Taylor, "Altruism in Viscous Populations—An Inclusive Fitness Model," *Evol Ecol* 6 (1992): 352–356; P. D. Taylor and S. A. Frank, "How to Make a Kin Selection Model," *J Theor Biol* 180 (1996): 27–37.

one could also say they were being altruistic because all will die if raided by a predator. Conversely, even in a large widely distributed population, the families can be thought of as groups within the whole population. If everyone of Smith lineage helps one another, and everyone of the Jones lineage does not, and the Smiths and Jones don't intermarry, then the population will come to consist of Smiths, and the Jones will dwindle away because the Smiths as a group out-produce the Joneses as a group. So, group selection can look like kin selection and kin selection can look like group selection, all depending on spin. Presently, therefore, many are debating what to make of this situation, should we throw away "group selection" in favor of "kin selection," or the reverse, or keep both names?[29]

The common element to both kin selection and group selection, regardless of spin, is that they both involve what is being called "multi-level selection," which means that selection is taking place in opposite directions at two levels simultaneously. Within a group or family, selection is favoring selfishness, and between groups or families, selection is favoring cooperation and altruism. Within a group of foragers, for example, selection tends to favor the individual who gobbles up as much food as possible while leaving others to starve, whereas between groups, selection favors the individual who shares food because if the group members aren't healthy enough, they will be unable to assist in predator protection leading to the death of the selfish individual. So, the overall

---

29. Benjamin Kerr and Peter Godfrey-Smith, "Individualist and Multi-level Perspectives on Selection in Structured Populations," *Biology Philosophy* 17 (2002): 477–517; Elliott Sober and David Sloan Wilson, "Perspectives and Parameterizations: Commentary on Benjamin Kerr and Peter Godfrey-Smith's 'Individualist and Multi-Level Perspectives on Selection in Structured Populations,'" *Biology Philosophy* 17 (2002): 529–537; Lee Alan Dugatkin, "Reply: Will Peace Follow?" *Biology Philosophy* 17 (2002): 519–522; Stuart A. West, Ashleigh S. Griffin, and Andy Gardner, "Evolutionary Explanations for Cooperation: Review," *Current Biology* 17 (2007): R661–R672; David Sloan Wilson and Edward O. Wilson, "Rethinking the Theoretical Foundation of Sociobiology," *Quarterly Review Biology.* 82 (2007): 327–348; S. A. West, A. S. Griffin, and A. Gardner, "Social Semantics: Altruism, Cooperation, Mutualism, Strong Reciprocity, and Group Selection," *J Evolutionary Biology* 20 (2007): 415–432; D. S. Wilson, "Social Semantics: Toward a Genuine Pluralism in the Study of Social Behaviour," *J Evolutionary Biology* 21 (2008): 368–373.

tendency to share that evolves will balance individual selection pushing for selfishness with group selection pushing for sharing. The same can be said if the group consists of families, the Smiths and Joneses, rather than a group of foragers. Individual selection favors sibling rivalry whereas group selection favors sibling cooperation. The spirit of multi-level selection then, is to visualize traits as the outcome of two selection processes, one physically nested within the other, with both operating simultaneously in opposite directions according to a common time scale measured in generations.

Our two-tier theory for the evolution of social behavior is completely different from multi-level selection because it is not about two levels of selection—no selection occurs at the lower tier, the only selection occurs at the higher tier. The two tiers are on different time scales, a within-generation time scale for the lower tier and a between-generation tier at the higher tier. Our lower tier models a social-developmental process, not a selection process. The phenotype that emerges from this social-development process earns a fitness accumulated throughout its life that is transferred to an individual-selection model for evolution at the higher tier. Our higher tier is ordinary individual-level selection.

If group selection and kin selection are in most senses equivalent, then they are equally common or rare. If group selection is indeed rare, as widely believed, then so is kin selection. If so, I suggest that many instances of cooperative behavior are best explained as the kind of team play envisioned in our behavioral tier, combined with ordinary individual-level natural selection in the evolutionary tier, and not as the result of multi-level selection, either kin or group.

NINE Family Harmony and Discord

Many readers will recall the American sitcom, *All in the Family*, featuring Archie Bunker, a comedy inspired by the BBC sitcom, *Til Death Us Do Part*. Although most commentators recall the Archie Bunker character primarily for its portrayal of bigotry, I most remember the continuing bickering between Archie and his wife, Edith. In one episode when Edith is entering menopause, Archie shouts, "If you're gonna have a change of life, you gotta do it right now. I'm gonna give you 30 seconds!"

For most sexual-selection advocates today, Archie and Edith Bunker's family dynamics amount to a romantic caricature of much nastier real-life dynamics found in animal families. For example, Geoff Parker recently wrote, "The family is now perceived as a cauldron of conflict, with each of the players having different interests . . . sexual conflict, parent-offspring conflict, and sib-competition simultaneously."[1] He's serious. I term this family-life-is-nothing-but-conflict view, the "Archie-Edith Bunker" hypothesis of animal family dynamics. Yet again, the question before us is not whether the Archie-Edith Bunker picture of family dynamics is appealing, repugnant, or even ludicrous in its hyperbole, but whether it is true.

Sexual-selection advocates are confident in their position. As we have seen before in Chapter 6, sexual-selection advocates attribute the universality of sexual conflict to the very origin of sperm and egg. Parker declares, "A primitive form of sexual conflict may have occurred during the early evolution of anisogamy."[2] But readers will also recall from Chapter 6 that the evolution of anisogamy can result from selection for an increased sperm-egg contact rate, and not necessarily from sperm-egg conflict, as sexual-selection advocates insist. Other writers even raise the ante on behalf of the universality of conflict in family dynamics, claiming "there has been a dramatic shift in the prevailing view of matings as being essentially 'a good thing' for both participants, to one in which they are regarded as 'bad' for females." Previous work is "mistakenly viewing male-female interactions as more benign than they actually are."[3] Well, after the bravado on behalf of universal family conflict has evaporated, we are left with the question, are animal families really a more devious and nasty counterpart of Archie and Edith Bunker?

Let's start then with who makes up a family, indeed whether any family exists at all. I consider a "family" to be a social group that produces offspring. I've previously reviewed in *Evolution's Rainbow* the many styles and organizations of families that occur among the vertebrates. Families include a monogamous male and female sharing a nest, extended families in which the offspring remain at the nest as helpers (what I term *post-zygotic* helpers), families in which helpers also participate

1. G. A. Parker, "Behavioural Ecology: The Science of Natural History," in *Essays on Animal Behaviour: Celebrating 50 Years of Animal Behaviour*, eds. J. R. Lucas and L. W. Simmons (New York: Academic Press, 2005), 23–56.

2. G. A. Parker, "Sexual Conflict over Mating and Fertilization: An Overview," *Phil Trans R Soc B* 361 (2006): 235–259.

3. T. Tregenza, N. Wedell, and T. Chapman, "Introduction. Sexual Conflict: A New Paradigm?" *Phil Trans R. Soc B* 361 (2006): 229–234.

prior to the birth of any offspring (what I term *pre-zygotic* helpers), families with one male and multiple females (polygyny) or one female and multiple males (polyandry), families with multiple types of males and multiple types of females, and of course, solitary parents who may or may not provide for their young. This list is abbreviated and all the arrangements are too numerous to detail. A family is a reproductive social group, a "firm" whose product is offspring, and it differs from other social groups, such as social foraging groups whose product is the capture of food, or social herds, schools, and flocks whose product is insurance and protection from predators.

Styles of family organization vary across the classes of vertebrates. As mentioned in Chapter 3, among fish species in which parental care is provided, usually the male provides the care, not the female.[4] A male contribution to parental care is common in birds.[5] Perhaps the most dramatic fact in family organization has been summarized by Clutton-Brock: "In contrast to birds, where over 90% of species are typically monogamous, the males of more than 90% of mammalian species are habitually polygynous."[6] This mammal-bird difference sets up what might be the most elemental question for the biology of family dynamics. As Clutton-Brock puts it, "Why, in some species, do males forego breeding polygynously to pair with a single female throughout most, or all of their lives?" Continuing, "Monogamy is one the most puzzling of mammalian mating systems, for it is not clear why males should confine themselves to breeding with a single female."

---

4. J.D. Reynolds, N.B. Goodwin, and R.P. Freckleton, "Evolutionary Transitions in Parental Care and Live Bearing in Vertebrates," *Phil Trans R Soc Lond B* 357 (2002): 269–281; cf. also Elizabet Fosgren et al., "Unusually Dynamic Sex Roles in a Fish," *Nature* 429 (2004): 551–554.

5. D. Lack, *Ecological Adaptations for Breeding in Birds* (London: Chapman & Hall, 1968); R. Pierotti and C.A. Annett, "Hybridization and Male Parental Investment in Birds," *Condor* 95 (1993): 670–679; M.C. McKilrick, "Phylogenetic Analysis of Avian Parental Care," *Auk* 109 (1993): 828; Ellen D. Ketterson and Val Nolan Jr, "Male Parental Behavior in Birds," *Annu Rev EcoL Syst* 25 (1994): 601–628; A. Cockburn, "Prevalence of Different Modes of Parental Care in Birds," *Proc Roy Soc B* 273 (2006): 1375–1383.

6. T. Clutton-Brock, "Review Lecture: Mammalian Mating Systems," *Proc R Soc Lond B* 236 (1989): 339–372.

Clutton-Brock's answer to why male birds usually contribute to off-spring care at the nest in contrast to mammals is that, for birds, "females are free to choose mating partners on the basis of phenotype or territory quality . . . models of this kind are of limited relevance to the majority of social mammals where females seldom disperse far from their natal area." That is, birds have more monogamy than mammals because flight endows female birds with more opportunity to choose their mates than female mammals have. Female birds can check out prospective husbands by flying from club to club around town, whereas a female mammal is stuck walking to the nearest block party. With so much choice before her, a female bird can demand a husband who is faithful and helps with the dishes at home, while a female mammal can't. Clutton-Brock's theory assumes that males inherently don't want to stick around and help with the young, can't stand doing the dishes at home, and must be manipulated to do so by a female's threat of turning to someone else. Clutton-Brock assumes males are inherently promiscuous, and their baser instincts must be thwarted by female manipulation.

Why should males inherently be promiscuous, as Clutton-Brock assumes? Because sperm are cheap and higher "mating success" is attained through abandoning the nest in endless search for new sex partners—the usual sexual selection narrative. Thus, Clutton-Brock, along with other sexual-selection advocates, regards male promiscuity as the original, or primitive, baseline condition, and monogamy as a secondary or derived specialization.

My position is just the opposite. I regard male parental care, including the monogamy it entails, as the original baseline condition and male promiscuity in mammals as a tactic of last resort, a derived specialization. I hypothesize that male promiscuity in mammals arises because female control of the young through internal fertilization, gestation, and subsequent lactation prevents male access to the offspring they've sired, denying them the opportunity to contribute to the survival of the young. Because mammalian males can't enhance their evolutionary success at home, they must settle for playing the field, the best of a bad situation. In birds, the young who reside in the nest as eggs or nestlings are accessible to both parents, who can then both contribute to the successful rearing of

Table 11 Hypothetical Payoff Matrices for Control of Nestlings

		Female	
		SHARE NESTLING ACCESS	CONTROL NESTLING ACCESS
Male	HELP AT NEST	(6, 6)	(1, 8)
	ABANDON NEST	(3, 3)	(2, 2)

the young. Avian males can enhance their evolutionary success by staying home and helping to guarantee the survival of the young they have sired.

The payoff matrix in Table 11 illustrates my hypothesis, which is located in the behavioral tier. The matrix pertains to a pair of birds in some habitat and describes their hypothetical capabilities at raising young depending on whether the male helps at the nest and depending on whether the female retains control of the young for herself.

Suppose that a male and female in the same vicinity start in the lower left, where a female lays, say four eggs, but must provision the nestlings herself and experience hazard during her foraging. This hazard reduces the expected contribution to her lifetime fitness from this season's work to a net three eggs. The male meanwhile, does not participate in the nesting activities and enjoys only the expected contribution to his lifetime fitness of the very same three eggs that the female has reared. So, the option exists for the male to help at the nest allowing the female to lay eight eggs. Because the female is also still foraging along with the male, the expected contribution to her lifetime fitness from the season's work is now a net six eggs, reflecting the joint work of both. The new total of six is also experienced by the male. The male then moves up to the top left position in the payoff matrix. From this point an option becomes available to the female who laid the eggs and who initially possesses them. She can exert control of the eggs for herself, forcing the male to do all the foraging. This action moves the pair to the top right of the payoff

Table 12   Control of Nestlings: Movement Diagram for Individual
Competitive Play

		Female		
		SHARE NESTLING ACCESS		CONTROL NESTLING ACCESS
	HELP AT NEST	↘	→	↙
Male		↑	NCS	↓
	ABANDON NEST	↗	←	↖

matrix. Here the female receives the full benefit of all eight eggs she lays
and because she is not incurring hazard during foraging, her fecundity
is not discounted by any risk of mortality. The male meanwhile incurs a
drop in net fitness to one egg, reflecting the great increase in foraging
hazard he now experiences. So at this point, the male abandons the nest,
which moves the pair down to the lower right position in the payoff matrix. In this position, the female now can rear only two eggs because she
has produced more than she can raise herself and the brood suffers
overall. Finally, the female reduces her fecundity back to the original
four eggs of which she can successfully rear a net three, after allowing
for the discount from the mortality risk she experiences during her
foraging.

Table 12 presents a movement diagram for the payoff matrix of Table
11, assuming the two birds are playing as individuals competitively
against each other. A mixed-strategy, Nash competitive equilibrium
(NCE), in this game occurs where the male independently helps at the
nest $\frac{1}{3}$ of the time and abandons the nest $\frac{2}{3}$ of the time, earning a fitness increment of $2\frac{1}{4}$ net eggs per season. Meanwhile, the female
shares nestling access $\frac{1}{4}$ of the time and retains control the eggs for herself $\frac{3}{4}$ of the time, earning a fitness increment of four net eggs per season. The individual-play dynamics in the vicinity of this NCE are cyclic
with a clockwise rotation, a trajectory flow indicated by the arrows in
Table 12.

Table 13   Control of Nestlings: Movement Diagram for Competitive Play

		Female	
		SHARE NESTLING ACCESS	CONTROL NESTLING ACCESS
Male	HELP AT NEST	NBS	←
	ABANDON NEST	↑	↖

Because the male earns about 2¼ net eggs per season at the NCE, he can initiate bargaining by abandoning the nest and hanging out. This drops the female's payoff from the four net eggs she was earning at the NCE down to three net eggs. They are now in the lower left corner of the payoff matrix. In light of this threat, it's advantageous for the female to bargain. The Nash bargaining solution (NBS) is the upper left corner at which both parties earn six net eggs per season. If the male and female birds then play cooperatively as a team, with coordinated actions taken to increase their team fitness function, their actions converge on the upper left position of the payoff matrix, as depicted in the movement diagram of Table 13.

According to this scheme, birds are more likely to attain a cooperative approach to offspring rearing (top left of payoff matrix) because the eggs contained in a nest are open to shared access by both the male and female, making the top left of the payoff matrix feasible for birds. In contrast, mammals are more likely to oscillate between female control of young and male abandonment of parental care (top right and lower left of payoff matrix), because mammalian young are relatively inaccessible to males, making shared access to the young impossible. In mammals with long life spans and long post-natal times to maturity, the young are increasingly open to shared access so that mammalian males can approach the degree of parental care readily possible in birds. Therefore, males in short-lived species should be more promiscuous than in long-lived species, all else being equal.

How does the hypothesis introduced in Tables 11 to 13 compare with the ideas in Clutton-Brock's review? Clutton-Brock's first hypothesis for the occurrence of monogamy in birds relative to mammals is that, because of the capability of flight, female birds have a vast choice of male partners compared to male mammals, as already mentioned. Clutton-Brock also offers two more possible explanations for monogamy.[7] His second hypothesis is a variant on the familiar story of males possessing as many females as possible. In special situations, the maximum a male can possess is only one female, a *harem of one*, as I term it. In some circumstances, particularly environments of low productivity, the defense of territories large enough to contain two females is impossible, or as Clutton-Brock puts it, 'it would be uneconomical for males to defend a larger area" than that which contains only one female. I agree this explanation is possible sometimes and have interpreted the monogamy we have studied in the smaller lizard on the Caribbean island of St. Martin as harem-of-one monogamy.[8]

Clutton-Brock's third hypothesis for monogamy initially seems similar to my own. He writes, "The most convincing explanations of monogamy are based on arguments that males can achieve a higher breeding success by guarding a single female and helping her rear their joint young than by attempting to breed polygynously. Two potential benefits of monogamy for males are that it increases the reproductive rate of their partners (and hence their own breeding success) as a consequence of the male's involvement in parental care, and that it increases the certainty of paternity, because males can accompany individual females." Clutton-Brock seemingly finds this situation to be unusual in mammals. In response, I observe that males who can achieve a higher breeding success by helping to rear the joint young are unusual in mammals compared to birds precisely because female mammals can

7. T. Clutton-Brock, "Review Lecture: Mammalian Mating Systems," *Proc R Soc Lond B* 236 (1989): 339–372.
8. H. Pereira, S. Loarie, and J. Roughgarden, "Monogamy, Polygyny and Interspecific Interactions in the Lizards *Anolis Gingivinus* and *Anolis Pogus,*" *Caribbean J Science* 38 (2002): 8–12.

monopolize the young more than female birds can. Therefore, male promiscuity in mammals emerges as an unintended consequence of female mammals' capacity for internal gestation and lactation, not as the expression of an inherent male characteristic.

Continuing, let's assume now that two birds at a nest have negotiated and come to a joint goal, such as that in the upper left corner of Table 11's payoff matrix. Both male and female are agreed to pursue their joint interest cooperatively. What behavior should we expect from them as the supply of worms, caterpillars, together with exposure to hawks, snakes, foxes, and other predators varies day by day. In biparental care, how should the birds carry out their joint purpose day by day as the nesting season progresses. This is family life for a monogamous pair of birds. Will it be the avian counterpart of Archie and Edith Bunker, or some more sophisticated family dynamic?

Sexual-selection advocates have developed a theory for avian family life based on assuming that male and female have different objectives. Starling parents and orange-tufted-sunbird parents quickly respond to each other when determining the amount of food each provides to their young.[9] To address the observations on starlings and sunbirds, McNamara, Houston and colleagues propose a model, hereafter called MH,[10] based on assuming that "a conflict of interest exists, with each parent preferring the other to work hard." They derive a mathematical formula to predict how one bird is supposed to change the amount of work it does in response to what the other is doing. For example, suppose one bird reduces its delivery of food by one item, then the other bird is supposed to compensate by increasing its delivery only a little bit, so as not to fully compensate, and thereby avoid being taken advantage of. Or conversely, if one bird "generously" adds an item to what it delivers the nestlings,

9. J. Wright and I. Cuthill, "Biparental Care: Short-term Manipulation of Partner Contribution and Brood Size in the Starling, *Sturmus Vulgaris*," *Behav Ecol* 1 (1990): 116–124; S. Markman, Y. YomTov, and J. Wright, "Male Parental Care in the Orange-tufted Sunbird–Behavioural Adjustments in Provisioning and Nest Guarding Effort," *Anim Behav* 50 (1995): 655–669.

10. J. McNamara, C. Gasson, and A. Houston, "Incorporating Rules for Responding into Evolutionary Games," *Nature* 401 (1999): 368–371.

the other bird should "selfishly" reduce its delivery by some small amount to take advantage of this unexpected gift. If both parties play back and forth with each other this way, each with their linear "best response" to the other, the amount that each delivers to the young will come to an equilibrium in behavioral time. This equilibrium is the NCE which results if each individually plays selfishly against the other in behavioral time based on male and female having different evolutionary objectives. Sexual-selection advocates expected smooth sailing as these predictions awaited confirmation.

A literature then developed to test such models of selfishness in biparental care. Schwagmeyer and coworkers[11] investigated the impact of lowering one parent's investment on the other's investment with house sparrows. They placed small lead fishing weights on males and on females to lower each's effectiveness in providing food to the nestlings. Remarkably, Schwagmeyer and coworkers conclude that "The apparent insensitivity of both males and females to changes in their mates' parental behavior, and the ineffectiveness of current partner behavior at predicting an individual's provisioning effort, fail to conform to assumptions of biparental care models that require facultative responses to partner deviations in effort."

More recently, Hinde[12] discovered even more problematic results. In her review of previous work, Hinde observes that various handicapping studies have resulted in "no increase," "partial compensation," and "full compensation" and other techniques have similarly yielded variable results. Her own study, therefore, introduced a new, nonhandicapping technique. She played tape recordings of begging calls to elicit increased provisioning using a British songbird. The result is particularly important: "Contrary to the predictions of partial compensation models, both sexes increased provisioning significantly when their partners increased provisioning in response to playback." She termed this kind of positive

11. P. Schwagmeyer, D. Mock, and G. Parker, "Biparental Care in House Sparrows: Negotiation or Sealed Bid," *Behav Ecol* 13 (2002): 713–721.

12. Camilla A. Hinde, "Negotiation Over Offspring Care?–A Positive Response to Partner-Provisioning Rate in Great Tits," *Behav Ecol* 17 (2006): 6–12.

response, *matching*. Now, this finding directly contradicts the selfish-nest-mates theory of avian biparental care and appears to be a serious refutation of that genre of models.

Hinde further noticed that some data in the paper by Schwagmeyer and coworkers also apparently showed matching behavior as well. That is, females weighted down with weights initially decreased provisioning but apparently learned to adapt to their handicap so well that they wound up increasing their overall provisioning to the nestlings. The males in turn increased their provisioning rate too, which Hinde comments as "a result which is intriguing and counter to previous predictions."

Johnson and Hinde then developed a work-around to the MH model to obtain better agreement with the data.[13] They retain the assumption that "each parent benefits if the other does more of the work involved in raising their offspring." What's new is that Johnson and Hinde, hereafter called JH, assume each partner has limited information about the status of the brood and must look to the amount of food that their nest mate brings to the brood as an indication of need. If a bird sees its partner bring more food to the nestlings, it can reassess its evaluation of how much food the brood needs and can then increase the amount of food it also brings the brood. This tendency to match the partner because of the information value to what the partner does is countered by the tendency to prefer the partner to do most of the work. The JH model therefore envisions a tension between selfishness and the need for information—selfishness leads a bird to counter its partner's effort and the need for information leads a bird to match for its partner's effort. Depending on the how much information the birds have about the needs of the brood, one or the other of these, selfishness or the need for information, may predominate, leading to a net outcome that can range from partial compensation, to no effect, to matching. Good information about the needs of the brood allows selfish compensation to predominate, whereas uncertainty about the needs of the brood leads matching to predominate.

13. Rufus A. Johnstone and Camilla A. Hinde, "Negotiation Over Offspring Care—How Should Parents Respond to Each Other's Efforts?" *Behav Ecol.* 17: 818–827.

I agree that the JH model offers a possible work-around to explain why data on nestling provisioning at bird nests contradicts the MH model for selfish nest mates. But is JH the correct explanation? Does the matching behavior that contradicts MH in fact result because the need for information about the brood's state trumps the selfishness assumed to underlie male-female dynamics? Alternatively, maybe the male and female are actually on the same page to begin with. As members of the same reproductive team, maybe they are cooperating and not working against each other. Maybe the varied responses to changes in each other's effort reveal a variety of ways to be most helpful to each other at increasing brood yield under varied circumstances.

As our lab group was discussing the selfish-nest-mates theory of biparental care, Erol Akçay suggested the need for a "null model" against which to assess putative evidence of family conflict. What, he asked, would be expected in experimental treatments such as handicapping birds or playing back nestling calls if the birds were perfectly cooperating? What would the birds do if, hypothetically, they were carrying out coordinated activities while pursuing a common goal of maximizing number of young successfully reared from their nest together? If the observed response of the birds to an experimental treatment is different from what would be expected if they were perfectly cooperating, only then should one begin to claim that evolutionary conflict exists in family life.

So, Erol developed a model for how the production of reared young from a nest depends on the time allocations of both the male and the female each to two activities, say, bringing food to the young, and patrolling for predators.[14] The formula that describes how many offspring are successfully reared depending on the fraction of the day the male and female each spend foraging and spend patrolling for predators is called a "production function" for the nest. It describes the common payoff to both depending on their joint activities and would be used to predict the

14. Erol Akçay and Joan Roughgarden, "The Perfect Family: Biparental Care in Animals," 2009: manuscript in preparation.

dual entries in the upper left corner of the payoff matrix in Table 11. The nest production function is an empirical quantity and should be measured in the field. Using the nest-production function, one can solve for the best joint allocation of effort into the two activities by the male and the female. We call a nest pair whose joint activities maximize a common nest production function, a "perfect family."

Using the common nest production function, one can also calculate how a perfect family will respond to an experimental treatment. To illustrate, suppose the female spends t_f feeding and $1 - t_f$ patrolling, and the male spends t_m feeding and $1 - t_m$ patrolling. Let's imagine two possibilities for how the male's and female's work can be combined to yield the nest's production.

One possibility is that the male and female each raise some young, which are then pooled to yield a total production. Suppose that the contribution to nestling rearing from the female is $t_f(1 - t_f)$ and that from the male is $t_m(1 - t_m)$. The young each rears for their common pool is the product each's foraging and patrolling times—the young die if not fed or if predators attack, so the product formula captures the need for both these activities. The overall nest production is then the sum of both the male and female contributions. In this case, the birds are pooling separately produced *products*. This sum is the nest production function. According to this formula for the nest production function, the number of young reared by the female is maximized when she spends 50% of her time foraging and 50% of her time patrolling, and similarly for the male.

Now suppose an experimental treatment handicaps the male, say by attaching a tiny lead weight to his body, and as a result, a male's foraging activity catches only one half of what it used to. Hence the value of the male's foraging time is now halved, and the number of young reared by the male drops to $(\frac{1}{2} t_m)(1 - t_m)$. Now this quantity, although necessarily less than what the female raises, is still maximized by the male putting 50% of his time into foraging and 50% of his time into patrolling. This is the best allocation of his effort, whether he is handicapped not. So, if the nest production is the sum of separate contributions, then the female's allocation should not change in response to handicapping the male.

The other possibility is to imagine that the male and female pool *effort* toward raising a common clutch of young, in which case the production would be $(t_f + t_m)((1 - t_f) + (1 - t_m))$. Here, they are pooling effort to produce a common product. Because the male and female efforts are interchangeable in the formula, all that matters is that the sum of t_f and t_m should equal 50% of their combined efforts. A split in which both do equal amounts of foraging and patrolling is just as good as a split in which one does all the foraging and the other does all the patrolling. As long as the total effort by both to foraging equals the total effort by both to patrolling, their common clutch size is maximized. Now if the male is handicapped, so that the nest production becomes $(t_f + \frac{1}{2} t_m)((1 - t_f) + (1 - t_m))$, the best joint allocation is total specialization, with the female doing all the foraging and the male doing none of the foraging—he does all the patrolling. In this case, the female overcompensates in response to the male's handicap.

Finally, let's imagine that the nest production function is a mixture of these two extremes. Perhaps the male bestows special attention on some of the nestlings, and the female bestows hers on the others, but they still tend all the nestlings to some degree. If the mixture between pooling privately tended young and jointly tended young is say, 50:50, then the nest production function becomes $0.5[t_f(1 - t_f) + t_m (1 - t_m)] + 0.5[(t_f + t_m)((1 - t_f) + (1 - t_m))]$. The best production is again realized when the female and male each spend 50% of their time foraging and 50% patrolling. But when the male is handicapped in this situation, the nest production function becomes $0.5[t_f (1 - t_f) + \frac{1}{2} t_m)(1 - t_m)] + 0.5 [(t_f + \frac{1}{2} t_m))((1 - t_f) + (1 - t_m))]$. Now, this production function turns out to be maximized when the female spends 65% of her time foraging and the male only 26% of his time foraging. Thus, the female has undercompensated in this case. The male has dropped his foraging from 50% of his time down to 26% because of the handicap whereas the female has increased from 50% only up to 65%.

The bottom line is that if the birds were, hypothetically, a perfect family in which both were adjusting their time allocations to maximize the number of joint young they successfully rear, then all possible responses to handicapping one of the parties can be realized depending on the characteristics of the nest production function. Thus, the experiments on biparental care in birds that refute the selfish nest-mate picture of avian monogamy are completely consistent with the perfect-family picture, depending on the nest production function.

Rather than asserting in advance that nest mates are necessarily selfish, with each preferring the other to do most of the work, sexual-selection advocates should demonstrate this assumption by falsifying a null model that postulates perfect cooperation between male and female. To the contrary, the data so far apparently falsify a selfish-nest-mates view of avian family life and instead seem consistent with the perfect-family view.

The perfect avian family is not necessarily lovey-dovey except during initial courtship. At the beginning of their relationship, negotiation and physical intimacy permit both the male and female bird to arrive at the same nest production function. To be a perfect family, it's sufficient for both birds' time allocations to jointly maximize this common nest production function. But the path whereby they come to realize the allocations that maximize their common production function might, or might not, involve continuing intimacy and team play. The joint maximization might also be realized through a happy confluence of decisions by each bird taken individually, given that both agree on what the production function is. Whether active cooperation and physical intimacy is present only initially or extends throughout the duration of their parenting depends on the details of the nest production function.

For example, let the nest production function be the pooled young brooded separately by the male and by the female. If each bird individually adjusts its foraging and patrolling time commitments to increase their shared production function, given the state of the other bird, then this sequence of individual adjustments leads to the NBS, even though they are playing independently. They both wind up at the NBS together simply because they share the same nest production function, and indeed, the NCE and NBS coincide.

Perhaps an analogy would be helpful. Imagine two hikers who wish to share a picnic at the top of a hill. If the hill is a simple dome, then regardless of where the hikers start, if each individually keeps climbing up, they both converge at the top, where they can sit down for lunch. Both are climbing the same hill, and no communication or cooperation is needed *en route* to arrive together at the top. The key is that they both agreed to climb the same hill; they don't also need to agree to climb it in synchrony.

On the other hand, recall the nest production function in which the birds pool effort on behalf of young brooded jointly by the male and female, and recall, too, that if the male is handicapped to reduce the effectiveness of his foraging effort, then the perfect-family response is for the female to do all the foraging and the male to do all the patrolling. Realizing this outcome may require coordination, however. If the birds specialize, one on patrolling and one on foraging, then either of the two possible complementary arrangements are NCEs. If each bird individually adjusts its time commitments to increase its production function, given the state of the other bird, the sequence of individual adjustments can lead to either the female doing all the foraging and the male all the patrolling, or vice versa, depending on who's doing what to begin with. Only one outcome is the NBS, however, which is for the female to do the foraging and the male the patrolling. The reverse is not as good. To be certain of attaining the NBS division of labor rather than converging on the suboptimal NCE, the birds should determine their time allocations through physically intimate team play.

Returning to the analogy, if the hill is not a dome, but has gullies and small promontories scattered about, a hiker programmed always to climb upwards can be diverted to a lesser lookout, while another winds up at the top of the hill. Instead, by communicating and comparing notes along the way, both hikers can meet for lunch at the top. Thus, whether coordination is needed *en route* depends on the topography of the production function, for a simple terrain, individual hill-climbing will yield the global optimum, whereas in complex terrain, team work will be needed to find the hill top, even if both hikers agree to climb the same hill.

Overall then, the perfect-family model offers a null hypothesis to contrast with the selfish-nest-mates model. In the perfect family, after the initial courtship, the ensuing interpersonal dynamics may be intimately cooperative or indifferently individualistic, but in either case, actions are taken to maximize a common nest production function.

Still, anyone who watches animals in nature sees instances of what must surely represent conflict—birds shrieking, lizards chasing, mammals head-butting and so forth. What then can account for instances of conflict which certainly exist sometimes? According to the sexual-selection advocates, conflict is everywhere and always coming first, whereas cooperation is uncommon and comes second. But according to social selection, reproductive social behavior begins in cooperation that may secondarily devolve into conflict. When does conflict rear its ugly head? Here's how these alternative hypotheses shape up.

Geoff Parker enunciates the sexual-conflict-is-everywhere position.[15] He writes, "Sexual conflict is a conflict between the evolutionary interests of individuals of the two sexes. Conflict requires . . . that the optimal outcomes for each sex cannot be achieved simultaneously." This definition seems to imply that sexual conflict is inevitable because no two individuals are likely to be able to simultaneously realize their evolutionary interests under all conditions. However, this definition confuses evolutionary with behavioral time scales. It may be to the advantage of two individuals to work as a team for a period of their lives. Their teamwork would be compromised by conflict during that behavioral period, irrespective of any possible differences in their long-range evolutionary objectives.

Readers of a certain age may recall their children, especially boys, playing a primitive computer game, *Dungeons and Dragons*. If one opens a door in the game and descends deeper into the dungeon, new dragons and other characters appear, all integrated into a local subplot. The dungeon of sexual conflict similarly has levels. Given that sexual-conflict is supposedly ubiquitous, one then opens a door into a subtheory to classify the types of conflict.

Parker writes, "unlike conflicts between non-reproducing individuals, with sexual conflict genomes may mix at fertilization to produce common progeny. Can this shared interest remove the conflict?" Parker's answer is no. The reasoning at this level of the dungeon revolves around the role of certain genes that may be passed from father to son. In the

15. G. A. Parker, "Sexual Conflict Over Mating and Fertilization: An Overview," *Phil Trans R Soc B* 361: 235–259.

father, these genes cause behavior that benefits him but harms the mother. Yet, these genes are ultimately supposed to be advantageous to the mother because her sons will similarly benefit by expressing the trait even though the female with whom her son mates is harmed just as she was. The mother is thus a genetic enabler of behavior that benefits males but is deleterious to females. This idea is called the "sons effect," or the "sexy-sons effect" in the sexual-conflict literature.

Parker writes, "coupling of male and female interests through the sons effect shows . . . three zones: (i) where the male trait is disadvantageous to both sexes and will not spread; (ii) the sexual conflict zone—where the trait is advantageous to males but disadvantageous to females (sexually antagonistic coevolution may occur . . . [for the female] to avoid mating with harmful males and/or to diminish harmful effects); (iii) the concurrence zone—where the trait is advantageous to both sexes (selection favours both the harmful trait in males and female traits to accept or prefer males with the trait)." Parker continues, "Most literature on sexual conflict assumes selection to be operating in the sexual conflict zone, both at a general level . . . and at the level of specific, harmful, male adaptations . . . and evidence that there has been female counter adaptation to such adaptations is sometimes available." In a particularly diabolical twist, Parker writes, "apparent female resistance against male persistence might represent a form of 'screening' where poor quality males (with low mating advantage) are rejected and high quality males eventually accepted."

Parker includes a taxonomy of how the phrase, "gain by losing" is used. "The term 'gain by losing' must be qualified. Eberhard[16] uses it in the context of gaining indirect benefits via sexy sons. Females may 'gain by losing' without the sons effect. It can be in female interests to acquiesce to mating simply because the direct costs of resistance are greater than the costs of allowing mating."[17] Thus, females gain by losing

16. W. G. Eberhard, "Evolutionary Conflicts of Interests: Are Female Sexual Decisions Different?" *American Naturalist* 165 (2005): S19–S25.

17. G. A. Parker, "The Reproductive Behaviour and the Nature of Sexual Selection in *Scatophaga Stercoraria* L. (Diptera: Scatophagidae) V. The Female's Behaviour at the Oviposition Site," *Behaviour* 37 (1970): 140–168; G. Arnqvist, "Multiple Mating in a Water Strider: Mutual Benefits or Intersexual Conflict?" *Anim Behav* 38 (1989): 749–756.

through either the "sexy-sons effect" or from what I term the "willing-victim effect."

Yet again, the issue before us is not whether one finds these thinly disguised rape narratives appealing or repugnant. The issue is whether a kind of rape actually does underlie all male and female relationships throughout nature. Sexual-conflict advocates do not acknowledge even the possibility of alternative hypotheses springing from a different point of view. Nonetheless, the scientific method requires alternative hypotheses.

According to the social-selection system, reproductive social behavior begins in cooperation. Conflict, should it occur, develops later. Some conflict arises when parties establish their threat points prior to reaching cooperative bargains. This conflict is transient, but possibly quite intense while it is taking place. If the two parties happen to be a male and female potentially sharing a common nest, then the conflict associated with establishing the threat points might be interpreted as sexual conflict, whereas the conflict would not be related to sex at all but would be an unremarkable labor dispute.

Another source of conflict might be called a "difference of opinion" about what would be needed to maximize a shared nest production function. Suppose for example, the female bird at a nest had happened to hear the cries of the nestlings begging for food. Meanwhile, suppose the male was out of calling range at the time, but happened to see some snakes and foxes lurking nearby. Based on their different experiences, the female bird might instinctively feel that the highest priority for maximizing nest production would be for the male to forage more and bring back food. Conversely, the male might feel instinctively that the highest priority for maximizing nest production would be for the female to keep a sharp eye out for predators, and not to take as many risks in foraging. So, each bird might have a different opinion on what should be done, even though they are agreed on working jointly to maximize their nest's offspring production.

A difference of opinion could play out in several ways. To illustrate, suppose the actual nest production function consists of pooling separately brooded eggs, which according to the previous formula is, $t_f(1 - t_m) + t_f(1 - t_m)$. Next, suppose the female's opinion of the nest production

Table 14   Perceived Payoff Matrix for Different Foraging Allocations of an
Almost-Perfect Family

| | | *Female* | |
		SHE WANTS 50%	HE WANTS 25%
Male	HE WANTS 50%	(0.5, 0.5)	(0.57476, 0.4375)
	SHE WANTS 50%	(0.4375, 0.57476)	(0.51226, 0.51226)

de-emphasizes the importance of male patrolling for predators and emphasizes his foraging effort according to the formula, $t_f(1 - t_f) + (t_m)^{3/4}$
$(1 - t_m)^{1/4}$. Accordingly, the female's opinion of what they should do to
maximize the offspring production from the nest works out to be for her
to spend 50% of her time foraging and for the male to increase his foraging time to 75% compared with the 50:50 split that each should have according to the actual production function. Conversely, suppose the
male's opinion of the nest production function de-emphasizes the female's foraging effort and emphasizes her patrolling effort according to
the formula, $t_f^{1/2}(1 - t_f)^{3/4} + t_m(1 - t_m)$. Accordingly, the male's opinion of
what they should do to maximize the offspring production from the nest
works out to be for her to reduce her foraging to 25% of her time, while
he continues with a 50:50 split by himself. Meanwhile, the actual production remains maximized if each carries out a 50:50 split of time allocated to foraging and patrolling.

We've called a family whose members have different opinions about
what their shared production function is, an "almost-perfect family." The
difference of opinion among family members can be represented as a
game which allows the possibility of compromise between family members. Table 14 offers the possibilities. The female can do what she wants
(forage 50% of the time) or what the male wants her to do (forage 25% of
the time). The male can do what he wants (forage 50% of the time) or
what the female wants him to do (forage 75% of the time). The payoffs to

Table 15 Realized Payoff Matrix for Different Foraging Allocations of an
Almost-Perfect Family

		Female	
		SHE WANTS 50%	HE WANTS 25%
Male	HE WANTS 50%	(0.5, 0.5)	(0.4375, 0.4375)
	SHE WANTS 75%	(0.4375, 0.4375)	(0.375, 0.375)

each combination of actions records each's opinion about the payoffs—this is the payoff that each perceives will result from their collective actions based on each's own experience. The top left represents the perceived payoffs if the female does what she wants and the male does what he wants. The bottom right represents the perceived payoffs if each does what the other wants.

By inspecting Table 14, one sees that the NCE lies at the upper left. If the birds act individually each merely does what it wants. On the other hand, the NCE is also the threat point, and the NBS is the lower right where each does what the other wants. So, if they play cooperatively as a team they wind up in the lower right. This outcome seems agreeable enough.

Alas, all is not sweetness and light. The opinions are incorrect. Neither the male nor female has correctly discerned what the other should do. The matrix of payoffs actually realized from these actions appears in Table 15. The table shows that the offspring produced if each acts individually exceeds the offspring produced if the male and female agreeably compromise with each based on incorrect opinions. This situation then leads to an Archie-Edith Bunker picture of family life. The parties are committed to a common nest production function, they try to get along with each other through team work, but cooperation doesn't work out. They are better off by playing as individuals, yet the bickering continues because the perception lingers for each that the other is not doing the right thing. This model predicts "sexual discord," as distinct from sexual conflict.

On the other hand, sexual harmony is also possible—when the realized payoff from the NBS happens to exceed the realized payoff from the NCE. Whether harmony or discord results in family dynamics depends on whether the opinions err in the same direction or different directions. Does compromising perceptions yield a decision closer to the objectively best course of action, or does compromising perceptions merely compound each's error? If the birds are lucky enough to have differences of opinion that compensate for each other's errors of assessment, then their almost-perfect family will enjoy sexual harmony. Alternatively, if the birds are unlucky enough to have differences of opinion that magnify each other's errors of assessment, then their almost-perfect family suffers continual sexual discord. Nonetheless, family discord emerges as a contingent secondary development, not an evolutionary necessity.

Future research should consider families larger than male-female pairs, and the possibility of hierarchy then becomes more important. Notions of staying incentives and profit sharing enter the picture and await further work that should draw on economic and management theory for firms called the "theory of teams" as cited previously in Chapter 9. Issues of honesty among nest mates and conditions that promote honest communication among team members about abilities and needs should be worked out. The key to allowing any forthcoming evolutionary theory of the family to be tested is to measure family production functions in the field so that the roles of the members can be predicted and compared with data.

TEN

Sharing Offspring with Neighbors

In popular culture, the robin has long been an icon for family love in nature, as a poem[1] called "Robin," taken from a contemporary website providing resources for school teachers, illustrates:

> Robin sits in the apple tree
> Singing sweetly down to me.
> She tells me of her cozy nest
> In the tree she likes the best.
>
> She tells me that her robin mate
> And she herself can hardly wait –
> They have four blue eggs, you see,
> That soon will be their family.

1. "Robin," available at http://www.canteach.ca/elementary/songspoems65.html

They'll live together in their cozy nest
In the tree that they love best.
Happy in her apple tree,
Robin sings her news to me!

Until rather recently, biologists shared this opinion of a robin's family life, albeit more prosaically. The ornithologist, David Lack, wrote in 1968 that "Well over nine-tenths of all passerine subfamilies are normally monogamous .... Polyandry is unknown."[2]

In the decades since then, the picture has changed. Although the poem doesn't say so explicitly, the robin's nest is understood to represent monogamy for the male and female in two senses—occupying a "cozy nest" together and also being the sole parents of the "four blue eggs" in their nest. But now it is clear that the nests of most birds contain some eggs parented by neighbors.

If two birds tend a nest together throughout a breeding season they are said to be "socially monogamous," or as I prefer, "economically monogamous." If all the eggs are parented by the two birds at the nest, then those birds are said to be "genetically monogamous." Separating economic monogamy from genetic monogamy avoids using contemporary Western heterosexual marriage as a metaphor for nature.

Some more terminology: The "pair-male" and "pair-female" are the birds sharing a nest—they are economically monogamous. "Extra-pair paternity" (EPP) refers to eggs whose male parent is a bird other than the pair-male, often a neighboring male. "Extra-pair maternity" (EPM) refers to eggs whose female parent is a bird other than the pair-female, again often a neighboring female. Together, these represent "extra-pair parentage" and result from "extra-pair copulations" (EPC) leading to "extra-pair fertilizations" (EPF). Also, EPY refers to "extra-pair young" in the nest.

Studies of extra-pair parentage are marred by value-loaded and pejorative language. Extra-pair maternity is called "egg dumping" or

---

2. D. Lack, *Ecological Adaptations for Breeding in Birds* (London: Methuen and Company, 1968).

intra-specific brood parasitism (IBP). EPP and EPM eggs are said to be "illegitimate" offspring, in contrast to the supposedly "legitimate" eggs whose parents are the pair-male and pair-female. Also, a pair-male whose nest contains EPP eggs is said to be "cuckolded," and their mother is said have "infidelity," as though they were married. This language precludes a theoretically neutral description of what is taking place.

The present-day picture of how parentage is distributed among economically monogamous pairs of nesting birds derives from modern molecular-genetic techniques for detecting paternity and maternity. A review in 2002 by Griffith and colleagues cited over 150 studies of avian parentage based on such methods.[3] The authors concluded that the average frequency of extra-pair offspring among economically monogamous bird species was 11.1% of offspring and 18.7% of broods. They also concluded that genetic monogamy (0% EPP) had been found in less than 25% of the economically monogamous bird species studied. Moreover, the authors observed that the "levels of EPP are often remarkably high within particular species, with a quarter of socially monogamous passerines having rates of EPP in excess of 25%," and that "among socially monogamous species, the most promiscuous bird detected to date is the reed bunting *Emberiza schoeniclus*, in which a recent study found that 55% of all offspring wesre fathered by extra-pair males and 86% of broods contained at least one chick fathered outside the pair bond."[4]

The discovery of extensive cross-parentage among bird nests caused a tizzy. Selfish-gene advocates rejoiced at pricking the bubble of romanticism surrounding nesting bird pairs and at data that seemingly confirmed their belief that female birds were deceiving their pair-male while surreptitiously following extra-pair males into the bushes.

The initial hypotheses for this cross-parentage suggested proximal ecological circumstances were responsible, namely (1) that either lots of

3. Simon C. Griffith, Ian P. F. Owens, and Katherine A. Thuman, "Extra-pair Paternity in Birds: A Review of Interspecific Variation and Adaptive Function," *Molecular Ecology* 11 (2002): 2195–2212.

4. A. Dixon et al., "Paternal Investment Inversely Related to Degree of Extra-pair Paternity in the Reed Bunting," *Nature* 371 (1994): 698–700.

birds were crammed together, making hank-panky inevitable in crowded quarters (high-breeding density hypothesis) or (2) that lots of birds were breeding at the same time leading to an avian counterpart of group sex (synchronous-breeding hypothesis). These hypotheses do not highlight any particular female agency in the mating that takes place— just lots of it going on and females get swept up in the activity. The 2002 review by Griffith and colleagues concluded that the high-breeding density hypotheses is probably not correct, and that the synchronous-breeding hypothesis remained unestablished even after much work intended to demonstrate it. They write, "There is little evidence that interspecific variation in the rate of EPP is due to variation in breeding density. If there is a relationship across species between breeding density and EPP then it is neither consistent nor strong." They continue, "Despite considerable empirical effort and much heated debate, it remains difficult to assess the role of variation in breeding synchrony in determining interspecific variation in EPP . . . [and] we feel it is too early to say that the breeding synchrony hypothesis is either important or trivial."

So, after reviewing hypotheses to explain avian cross-parentage in terms of immediate ecological circumstances, Griffith and colleagues turned to hypotheses that are extensions of the sexual-selection narrative. They write, "The main types of explanation for why females may seek EPP for their offspring . . . mirror the hypotheses that have been proposed to explain the evolution of secondary sexual ornaments in birds." These hypotheses do attribute agency to female choice. EPP is the outcome of a female's invitation to mate. The hypotheses then fall into three categories: (1) females solicit extra-pair males as insurance against possible infertility of their pair-male, (2) females solicit extra-pair males to get better genes for their offspring (indirect benefit), and (3) females solicit extra-pair males for a direct payoff, such as a "gift" of food or protection from predators (direct benefit). These classes of hypotheses further split into subhypotheses. For genetic benefits, a female might seek a male with better genes than her pair-male (genetic upgrade), seek a male genetically compatible to her own genes (genetic compatibility), or seek multiple males to yield a genetically diverse group of offspring (genetic diversity). After reviewing all the studies, Griffith and colleagues conclude, "only the 'good genes' [genetic upgrade] and 'genetic compatibility' hypotheses have received robust empirical support. It remains to be shown whether these are the general explanation for EPP in birds."

Subsequent work has not sustained this hope that the genetic upgrade and genetic compatibility hypotheses are general explanations for EPP in birds. As mentioned previously in Chapter 3, Erol Akçay has looked into the evidence that genetic benefits of any sort, good genes or compatible genes are involved in EPP.[5] He reviewed almost all published studies testing for genetic benefits from 1980 onwards, for a total of 121 papers pertaining to 55 species. To understand how the published studies obtained their data, let's consider exactly how the hypotheses are stated, following Erol's paper closely.

According to the good-genes (genetic-upgrade) hypothesis,[6] some males have "better" genetic quality than others. Males advertise their genetic quality and females prefer copulating with higher quality males. However, not all females can pair with the most preferred males, so females paired to nonpreferred males try to "trade up"—that is, have copulations with a more attractive male to obtain better genes for her offspring. In this way, females secure both the direct benefits of parental care from their pair-male as well as the genetic benefits of "good genes" from her extra-pair male. Males advertise their genetic "quality" by some signalling mechanism, which might be behavioral, such as audible volume or duration of a call, or morphological, such as bright feathers or a large body. Hence, good-genes studies look for contrast between the behavioral and morphological traits of the pair-male and the extra-pair male. Studies might also compare the traits in extra-pair offspring with traits in offspring sired by the pair-male with the same female. And in particular, extra-pair young should have higher fitness than young sired with the pair-male, assuming the same mother.

5. Erol Akçay and Joan Roughgarden, "Extra-pair Paternity in Birds: Review of the Genetic Benefits," *Evolutionary Ecology Research* 9 (2007): 855–868.

6. M.D. Jennions and M. Petrie, "Why do Females Mate Multiply? A Review of the Genetic Benefits," *Biol Rev* 75 (2000): 21–64; B.D. Neff and T.E. Pitcher, "Genetic Quality and Sexual Selection: An Integrated Framework for Good Genes and Compatible Genes," *Molec Ecol* 14 (2005): 19–38.

An alternative to the good-genes hypothesis is that each female prefers a different male, rather than all preferring the same male. Such differences in preference can be due to genetic incompatibility or high relatedness associated with inbreeding depression.[7] In this scenario, genetic benefits of EPC to females trace to the interaction between maternal and paternal genomic contributions. Females are predicted to pursue extra-pair copulations to increase the chances of securing a paternal contribution that is more compatible with their own contribution. In practice, most studies of the compatible genes hypothesis use relatedness or genetic similarity between partners as the test variable. Empirical tests of the compatible genes hypothesis are similar to those of the good genes hypothesis in that extra-pair offspring should have higher fitness than within-pair offspring. A difference between good-genes and compatible-genes hypotheses however, is that the compatible-genes hypothesis is symmetric with respect to sex. In the good-genes hypothesis, the males carry the good genes. In the compatible-genes hypothesis, neither sex has the better genes, but both maternal and paternal genomes must be biochemically agreeable.

Erol searched three databases (Biosis, ISI Science Citation Index, and ISI Social Science Citation Index) using keywords related to extra-pair paternity and genetic quality. He included years from 1980 to 2007 and selected studies that satisfied these criteria: (1) the study reported molecular exclusion and/or assignment of paternity; (2) the rate of extra-pair paternity was non-zero; and (3) the study presented at least one test of either of the two genetic-benefits hypotheses. Erol also supplemented this database with any earlier studies satisfying these criteria that had been cited in previous reviews.[8] In total, 121 studies were included in

7. J. A. Zeh and D. W. Zeh, "The Evolution of Polyandry I: Intragenomic Conflict and Genetic Incompatibility," *Proc R Soc Lond B* 263 (1996): 1711–1717; J. A. Zeh and D. W. Zeh, "The Evolution of Polyandry II: Post-copulatory Defences Against Genetic Incompatibility," *Proc R Soc Lond B* 264 (1997): 69–75.

8. A. P. Møller and P. Ninni, "Sperm Competition and Sexual Selection: A Meta-analysis of Paternity Studies of Birds," *Behav Ecol Sociobiol* 43 (1998): 345–358; S. C. Griffith, I. P. F. Owens, and K. A. Thuman, "Extra Pair Paternity in Birds: A Review of Interspecific Variation and Adaptive Function," *Molec Ecol* 11 (2002): 2195–2212.

the resulting database which is available online.[9] Erol located 121 studies conducted on 55 different species. Some species and genera were more heavily represented than others because researchers focus on favorite organisms.

The good-genes hypothesis was tested in 106 of the papers representing 51 species. For 32 species, the good genes hypothesis was tested with only one paper, whereas for 19 species, more than one paper was available. Here is the overall pattern: only 45 papers (42%), representing 22 species (43%) reported support for the good-genes hypothesis, and 61 papers (58%) representing 29 species (57%) reported no support for the good-genes hypothesis to explain EPCs.

The compatible genes hypothesis was tested in 34 of the papers representing 24 species. Here is the overall pattern: only 15 papers (44%) representing 11 species (44%) report support for the compatible-genes hypothesis whereas 19 papers (56%) representing 11 other species (44%) report no support for the compatible-genes hypothesis to explain EPCs. Two of the species (12%) had mixed results across multiple papers.

What are we to conclude about these results? The most conservative conclusion is that no conclusion can be made. Some sexual-selection advocates have claimed to us that these studies "may not have been able to test the appropriate hypotheses adequately." In my opinion, this response is obstinate and evasive.

Another position is to argue that the collective results actually support the good genes hypothesis. If we had 100 studies that test a hypothesis at the 0.05 significance level, and the good-genes hypothesis was not true, we would expect only 5 of them to show significant results. Therefore, the presence of 42% of papers showing significant effects might be interpreted as actually supporting the good genes hypothesis.

But that conclusion begs the question of what the null hypothesis is. The null hypothesis is not that traits believed to advertise good genes are never associated with the birds chosen by females for EPCs. The correct null hypothesis is that such traits are irrelevant to female choice of EPC

9. Online Appendix at http://evolutionary-ecology.com/data/2203appendix.pdf

partners. By this null hypothesis, whatever females are choosing for will, by coincidence, sometimes be associated with traits like colorful tails and sometimes not. Therefore, around 50% of the time female choice will line up with traits like colorful tails, and 50% not. The collective results do not differ from this correct null hypothesis.

Still another response by sexual-selection advocates is to argue that the good genes hypothesis need not explain extra-pair paternity in all species, it need explain EPP in only some species. But why? To be complete and testable, this argument would then need some super-theory that specifies which species to expect the good genes hypothesis to be important in, and which to expect good genes not to be important in. Not only does such a super-theory not exist, it would fly in the face of the rationale for the good-genes hypothesis to begin with. The good-genes hypothesis postulates that *all* species should have the problem of accumulated weakly deleterious mutations that females must avoid through their mate choice. In fact, for the good-genes hypothesis to be true, it must be true for all species, not just for some. Evidently though, the good-genes hypothesis is invalid as a universal hypothesis, and the 43% of species in which the good-genes hypothesis seems to be correct is coincidence.

We also note that in species where some evidence supports the good genes hypothesis, different populations and closely related species report evidence against it. Blue tits are a case in point. The relationship found in one population[10] is not found in others.[11] Contradictory relations between paternity loss and gain and male ornamentation have

10. B. Kempenaers et al., "Extra-pair Paternity Results from Female Preference for High-quality Males in the Blue Tit," *Nature (Lond)* 357 (1992): 494–496; B. Kempenaers, G. R. Verheyren, and A. A. Dhondt, "Extrapair Paternity in the Blue Tit Parus Caeruleus: Female Choice, Male Characteristics, and Offspring Quality," *Behav Ecol* 8 (1997): 481–492.

11. C. Krokene et al., "The Function of Extrapair Paternity in Blue Tits and Great Tits: Good Genes or Fertility Insurance?" *Behav Ecol* 9 (1998): 649–656; A. Charmantier et al., "Do Extra-pair Paternities Provide Genetic Benefits for Female Blue Tits *Parus Caeruleus*?" *J Avian Biol* 35 (2004): 524–532; A. Poesel et al., "Early Birds are Sexy: Male Age, Dawn Song and Extrapair Paternity in Blue Tits, *Cyanista* (Formerly *Parus*) *Caeruleus*," *Anim Behav* 72 (2006): 531–538.

been documented in the same population.[12] Finally, no evidence in support of the good genes hypothesis was found in closely related species, the great tit[13] and the coal tit,[14] despite comparable effort and similar or higher rates of extra-pair paternity. Such conflicting results support the view that "good genes" are irrelevant to female choice of EPP partners.

Furthermore, the studies collectively demonstrate no difference in the survival of extra-pair young and within-pair young. Most formulations of the good-genes hypothesis postulate genetic viability benefits to the offspring. The empirical studies, however, do not support this assumption, based on direct measurements of offspring viability. The good genes effect might also operate through increased attractiveness, rather than viability of offspring. The only study accounting for attractiveness as well as viability found no significant differences between number of grandchildren produced to the female through extra-pair young and within-pair young,[15] casting doubt on this version of the good-genes hypothesis as well.

Moreover, any good-genes theory intended to account for mate choice in extra-pair matings suffers from the familiar logical flaws of sexual-selection theory discussed previously in Chapter 3, namely, the paradox of the lek and the limited scope for detecting genetic quality differences

12. K. Delhey et al., "Paternity Analysis Reveals Opposing Selection Pressures on Crown Coloration in the Blue Tit (*Parus Caeruleus*)," *Proc R Soc Lond B* 270 (2003): 2057–2063; K. Delhey et al., "Fertilization Success and UV Ornamentation in Blue Tits *Cyanistes Caeruleus*: Correlational and Experimental Evidence," *Behav Ecol* 18 (2007): 399–409.

13. C. Krokene et al., "The Function of Extrapair Paternity in Blue Tits and Great Tits: Good Genes or Fertility Insurance?" *Behav Ecol* 9 (1998): 649–656; T. Lubjuhn et al., "Extrapair Paternity in Great Tits *Parus Major*—A Long Term Study," *Behaviour* 136 (1999): 1157–1172.

14. V. Dietrich et al., "Cuckoldry and Recapture Probability of Adult Males are not Related in the Socially Monogamous Coal Tit (*Parus Ater*)," *J Ornithol* 145 (2004): 327–333; T. Schmoll et al., "Long-term Fitness Consequences of Female Extra-pair Matings in a Socially Monogamous Passerine," *Proc R Soc Lond B* 270 (2003): 259–264; T. Schmoll et al., "Paternal Genetic Effects on Offspring Fitness are Context Dependent Within the Extrapair Mating System of a Socially Monogamous Passerine," *Evolution* 59 (2005): 645–657.

15. T. Schmoll et al., "Paternal Genetic Effects on Offspring Fitness are Context Dependent Within the Extrapair Mating System of a Socially Monogamous Passerine," *Evolution* 59 (2005): 645–657.

tracing to accumulated weakly deleterious mutations. Sexual selection is incorrect for within-pair female choice and also for extra-pair female choice.

Finally, using statistical methods called meta-analysis, Erol detected significant publication bias in the test for secondary sexual traits. Bias favored studies claiming to support a sexual-selection-like importance for secondary sexual traits and bias discriminated against results that undermine the sexual-selection-like importance of secondary sexual characters.

Overall, the data offer a devastating refutation of the good-genes hypothesis for female choice of EPP partners, a hypothesis that derives from a sexual-selection picture of reproductive social behavior.

Does the compatible-genes hypothesis fare any better? No. The collective results for the compatible-genes hypothesis are the same as for the good genes hypothesis; therefore, the same possible interpretations as for the good-genes hypothesis apply. Despite some recent prominent reports,[16] genetic similarity between socially paired individuals seems not to play much role in female choice of EPP partners.

Erol's findings that discount the importance of the genetic-upgrade and genetic-compatibility hypotheses in female choice of EPP partners agrees with the conclusion from a more limited survey by Arnqvist and Kirkpatrick in 2005 (hereafter abbreviated as AK).[17] However, rather than question sexual-selection theory, AK bestir the dragon of sexual conflict. If sexual selection isn't true, then by golly, sexual conflict must somehow be true instead.

AK argue that EPPs represent the outcome of male harassment, not female choice. Their argument begins from the premise that "females may suffer direct costs of infidelity, primarily in terms of reduced paternal care of offspring by their social mate." AK reviewed 12 species seeking a

16. D. Blomqvist et al., "Genetic Similarity Between Mates and Extra-pair Parentage in Three Species of Shorebirds," *Nature* 419 (2002): 613–615.

17. G. Arnqvist and M. Kirkpatrick, "The Evolution of Infidelity in Socially Monogamous Passerines: The Strength of Direct and Indirect Selection on Extrapair Copulation Behavior in Females," *American Naturalist* 165 (suppl): S26–S37 (2002).

statistical relation between parental care and EPPs. They presented a negative relationship overall between the proportion of the total parental care provided by the pair male *vs* the proportion of EPY in the nest. That is, statistically, the more extra-pair young in the nest, the lower the amount of time spent by the pair-male in parental care. They then argue that this reduction of pair-male time at the nest should be considered as lowering the female's fitness by reducing the total production from "her" nest. This putative reduction in her fitness is construed as "direct selection" against her for her "infidelities."

Using a single-tier evolutionary framework in which both EPCs and parental care are viewed as evolved traits, AK compared the indirect fitness benefit to a female that supposedly comes from an EPC genetic upgrade with the direct fitness cost stemming from the diminished parental care provided by the pair-male. Based on this comparison, they observe that "indirect selection is not generally present" and if it is, "it should be overwhelmed by the effects of direct negative selection". They state, "Indirect selection on EPC behavior in females is generally insignificant in comparison with direct selection." They conclude, "Indirect genetic benefits to offspring are unlikely to provide a general explanation for the evolutionary maintenance of EPC behavior in socially monogamous passerines because such effects are very weak at most and are overwhelmed by negative direct selection" and instead, "EPCs primarily reflect antagonistic coevolution between offensive male adaptations to gain extrapair paternity on the one hand and resistance adaptations in females and defensive adaptations in males on the other."

And yet again, the issue before us is not whether the claim is appealing or repugnant that EPPs are forced upon females to their detriment, but whether this claim is true, not just occasionally, but in general and necessarily so. I think the single-tier mathematical approach by AK is incorrect to begin with because it views EPCs and parental care as genetically determined traits rather than as tactics that emerge behaviorally through experience in local circumstance. The AK calculations are not "meaningful empirical estimates" as they claim, but a mathematical fiction.

In 2007, Griffith published an important critique of AK, and I high-light some of his criticisms here.[18] For starters, Griffiths points out that AK's data pertain to the results of behavior, not to behavior itself.

Mating, as we all know, does not necessarily result in fertilization and paternity. But AK continually frame their conclusions in terms of how fe-males behave as though there were a simple link between paternity and mating. Griffith writes, "More than 10 years ago, Dunn and Lifjeld[19] demonstrated convincingly that there was not a linear relationship be-tween the observed rate of EPCs and EPP across a number of species, and to this date, there remains little or no evidence to support a simple relationship between the two."

Griffith criticizes AK, saying "not a single one of the studies on which their analysis is based has ever investigated female copulation behavior directly." Griffith continues, "Naturally, by detecting the broods in which EPP occurs, these studies can identify females that definitely must have had at least one EPC. However, there will be no forensic trace of any number of EPCs that females may have had that did not result in EPP."

Griffith goes on to cite examples. On one hand, EPPs may greatly ex-ceed the number of EPCs that behavioral studies discern. He writes, "In the fairy wren *Malarus cyaneus*, more than 70% of offspring are sired by extragroup males,[20] and yet no one has directly observed an EPC in this species despite intensive behavioral study for more than 10 years."[21]

On the other hand, the EPPs may also be much less than the number of EPCs that behavioral studies discern. Griffith mentions the collared fly-catcher in which EPP is found to occur regularly in about 33% of broods.[22]

---

18. S.C. Griffith, "The Evolution of Infidelity in Socially Monogamous Passerines: Neglected Components of Direct and Indirect Selection," *Am Nat* 169 (2007): 274–281.

19. P. Dunn and A. Cockburn, "Extrapair Mate Choice and Honest Signaling in Coop-eratively Breeding Superb Fairy-Wrens," *Evolution* 53 (1999): 938–946.

20. R.A. Mulder et al., 1994. Helpers liberate female fairy-wrens from constraints on extra-pair mate choice. *Proceed Royal Society B Biol Sci* 255 (1994): 223–229.

21. M. Double and A. Cockburn, "Pre-dawn Infidelity: Females Control Extra-pair Mating in Superb Fairy-Wrens," *Proceed Royal Society B Biol Sci* 267 (2000): 465–470.

22. B.C. Sheldon and H. Ellegren, "Sexual Selection Resulting from Extra-pair Pater-nity in Collared Flycatchers," *Animal Behaviour* 57 (1999): 285–298.

But experiments[23] showed "that under normal circumstances, approximately 80% of female flycatchers are having EPCs but that generally these are not detected." Thus, "to date, in the only species in which we can make an estimate about the relationship between extrapair behavior and actual EPP, it seems that about 80% of females may be having EPCs and yet only about 30% of females normally have EPP in their broods." Griffith concludes, "Therefore, when we detect extrapair offspring in the broods of a sample of females in a population (e.g., 30%), we are unable to conclude that these are the only 30% of females in the population that have had EPCs. In fact, it is quite possible that all females may have had EPCs and that in only 30% of them did the sperm of extrapair sires achieve fertilizations."

The disconnect between mating and fertilization is important not only as a methodological criticism of AK's analysis, but for a substantive reason as well. Is the biological function of mating primarily to bring about fertilizations, or to achieve social outcomes, or some mixture of both? Recall the European oystercatcher in which some of the breeding arrangements contain two competing females in separate nests with the male splitting time between them, and other breeding arrangements that consist of two cooperating females in one nest with a male.[24] The competing females don't mate with each other, whereas the cooperating females do frequently copulate with each other, which obviously don't lead to fertilizations, but do contribute to their social infrastructure. Thus, AK's use of paternity analysis data to draw conclusions about female behavior is not merely technically inaccurate, but misunderstands the function of much, and possibly most, mating behavior.

23. G. Michl et al., "Experimental Analysis of Sperm Competition Mechanisms in a Wild Bird Population," *Proceed Nat Acad Sci USA* 99 (2002): 5466–5470.

24. M. P. Harris, "Territory Limiting the Size of the Breeding Population of the Oster-catcher (*Haematopus Ostralegus*)—A Removal Experiment," *J Anim Ecol* 39 (1970): 707–713; B. J. Ens et al., Life History Decisions during the Breeding Season, in *The Ostercatcher: From Individuals to Populations*, ed. J. D. Goss-Custard (Oxford: Oxford University Press, 1996), 186–218; D. Heg and R. van Treuren, "Female-Female Cooperation in Polygynous Oster-catchers," *Nature* 391 (1998): 687–691.

Griffith's second criticism of AK questions whether males actually do reduce their parental care in response to female "infidelity." Two types of data might be offered to support this assumption, data from comparisons across species, and data from within species. The possibility of a relation between a species-average degree of EPP and its average degree of male parental care has been disputed. A study in 1993 claimed that parental care drops in response to decreased certainty of paternity.[25] This claim was then challenged in 1999 based on reanalyzing the data plus additional data.[26] This challenge was then followed by the strong assertion that the relation does exist at all.[27] Studies on particular species also suggest the absence of any relation between male parental care and assurance of paternity. Wagner[28] writes of the razorbill, "On average, males contributed approximately equally to their mates in chick feedings and overnight nest attendance. There was, however, marked variation in relative male effort, with the proportion of male feeding relative to their mates' feeding ranging from 16 to 72%. No significant portion of this variation was explained by direct and indirect measures of males' confidence of paternity." The most recent studies continue in this vein.[29] Evidently, the very existence of a relation between EPCs and parental care is far from established. One review of this relation ends up dismissing the whole matter anyway by saying, "it is debatable whether the comparative analyses are at all relevant to the question of whether males adjust paternal investment in response to certainty of paternity."[30] Griffith agrees, "The association between EPP and paternal care across species is

25. A. P. Møller and T. R. Birkhead, "Certainty of Paternity Covaries with Paternal Care in Birds," *Behav Ecol Sociobiol* 33 (1993): 261–268.

26. P. Schwagmeyer et al., "Species Differences in Male Parental Care in Birds: A Reexamination of Correlates with Paternity," *Auk* 116 (1999): 487–503.

27. A. P. Møller and J. J. Cuervo, "The Evolution of Paternity and Paternal Care in Birds," *Behav Ecol* 11 (2000): 472–485

28. Richard H. Wagner, "Confidence of Paternity and Parental Effort in Razorbills," *Auk* 109 (1992): 556–562.

29. Seppo Rytkönen et al., "Intensity of Nest Defence is Not Related to Degree of Paternity in the Willow Tit *Parus Montanus*," *J Avian Biol* 38:273–277.

30. B. C. Sheldon, "Relating Paternity to Paternal Care," *Philos Transact Royal Soc B Biol Sci* 357 (2002): 341–350.

not relevant to the issue of whether individual males facultatively adjust parental investment in response to variable certainty of paternity."

Then Griffith states what I find to be the most important criticism of the idea that males withdraw parental care in response to female "infidelity." Griffith writes, "This idea has a number of logical problems, the most significant being that in socially monogamous birds there is no evidence that birds can discriminate between their own and extrapair offspring.[31] This means that if males significantly reduce care to the brood, as assumed by Arnqvist and Kirkpatrick,[32] they will harm their own offspring in addition to the extrapair offspring. It is difficult to produce a sensible model in which such a facultative reduction in paternal care can ever 'pay,' given that cases of males being completely cuckolded are exceptionally rare and that on average most of a male's genetic offspring are in his own nest rather than elsewhere."[33] So, the male is hurting himself as much as hurting the female by withdrawing his contribution to care of the nestlings, and it doesn't make sense for him to do this. Even if the male had sired only one of the many eggs in his nest, unless he's sired more eggs in some other nest to which he can proffer his care instead, it makes no sense to withdraw his care of the brood containing that one egg.

A study in 2005[34] claims to confirm AK by stating that "costs to females resulting from reduced parental care from cheated males constrain promiscuity," that "females exert resistance over EPFs when the costs of infidelity are high and, conversely, that the rate of EPFs increases when . . . costs of infidelity are low," and that "EPFs occur whenever females can expect a low fitness penalty from cheated males in response

---

31. B. Kempenaers and B. C. Sheldon, "Why do Male Birds Not Discriminate Between Their Own and Extra-pair Offspring?" *Ani Behav* 51 (1996): 1165–1117.

32. G. Arnqvist and M. Kirkpatrick, "The Evolution of Infidelity in Socially Monogamous Passerines: The Strength of Direct and Indirect Selection on Extrapair Copulation Behavior in Females," *Am Nat* 165 (suppl): S26–S37 (2005).

33. L. A. Whittingham and P. O. Dunn, "Effects of Extra-pair and Within-pair Reproductive Success on the Opportunity for Selection in Birds," *Behav Ecol* 16 (2005): 138–144.

34. Tomáš Albrecht, Jakub Kreisinger, and Jaroslav Piálek, "The Strength of Direct Selection Against Female Promiscuity is Associated with Rates of Extrapair Fertilizations in Socially Monogamous Songbirds," *Am Nat* 167: (2006): 739–744.

to their infidelity." The authors "conclude that, despite selection acting on males to achieve EPF and to protect within-pair paternity, species-specific rates of EPFs could be viewed as a function of fitness costs to females (resulting from the species-specific ability of males to control and punish female infidelity) and (but not necessarily) fitness gains to females from EPFs." Thus, species with high proportions of EPFs in the nests are species in which the males are unable for some reason to punish the females for their infidelities nor are they able to successfully guard their own paternity, causing the females to succumb to the EPFs forced on them. Hence, "males benefit from EPFs, but females adopt a 'best of a bad job' strategy when engaging in EPFs." This argument doesn't make sense. A strange asymmetry in male capability is being postulated—high-EPF species are those for some reason with males unable to punish infidelity in the pair-female and unable to defend their own paternity and yet able to coerce infidelity among extra-pair females.

To the contrary, females are well known to actively solicit EPCs. For instance, consider Razorbills, of the North Atlantic.[35] Males and females have the same color and overall shape as each other, and live in pairs at nests in a colony. A pair provides joint parental care for one egg laid each year. Most mating occurs in openly visible areas called arenas near the colony, even matings between the pair-female and pair-male. Indeed, 75% of the within-pair matings take place in the arena, even though a pair shares a nest together in the colony. Eight-seven percent of the extra-pair matings also occur in the arenas. Furthermore, males solicit EPC's with one another, not only with females. 41% of all extra-pair matings are between males, which is 18% of *all* mountings, including the within-pair male/female matings. Nearly $\frac{2}{3}$ of all males mount other males (an average of five partners apiece and as many as 16) and more than 90% of the males receive mounts from other males. Thus, extra-pair copulation effort by males is allocated somehow across both males and females.[36]

35. R. Wagner, "The Pursuit of Extra-pair Copulations by Monogamous Female Razorbills: How do Females Benefit?" *Behav Ecol Sociobiol* 29 (1992): 455–464.

36. R.H. Wagner, "Male-male Mountings by a Sexually Monomorphic Bird: Mistaken Identity or Fighting Tactic?" *J Avian Biol* 27 (1996): 209–214.

Here's a description by Wagner of the natural history of female choice in razorbills:[37] "females clearly controlled the success of copulations ... female razorbills have relatively long, stiff tails which they must lift for the male to achieve cloacal contact; in over 600 EPC attempts males did not appear to be capable of forcing females to lift their tails." Wagner further notes, "males often desisted from mounting females who had repeatedly rejected cloacal contact, implying that by mounting females, males had been testing the willingness of females to accept an EPC, rather than attempting to copulate by force" and "females often remained in place after resisting an EPC attempt, rather than fleeing or moving away once the male had dismounted; this permitted the same male to make further attempts, implying that females were inviting the mountings." Wagner concludes "females can not only control whether cloacal contact is achieved during mountings, but can also influence whether males mount them at all."

AK are aware that if females actively solicit EPCs, the entire premise that EPCs are forced upon them is dubious. So, AK downplay the natural-history evidence, calling it "rather anecdotal in most cases." They then conjecture that "females are in some way constrained from evolving resistance to EPCs such that EPCs might be the 'best of a bad job' for females," as though females are better off to volunteer for rape rather than suffer the real thing. AK write, "This does not in any way assume that overt coercion is involved in EPCs but rather that EPCs might result from a form of sexually antagonistic 'seduction.'" This is the willing-victim dime-story again offering theoretical cover for contrary data. Sexual conflict has become sexual selection's ether—undetectable but everywhere.

As our lab group considered the evident inadequacy of sexual-selection and sexual-conflict hypotheses to account for EPP, it became clear that some completely new thinking was needed. But how? Help arrived from an unexpected source—one of the responses to our *Science* article.[38]

37. R. Wagner, "The Pursuit of Extra-pair Copulations by Monogamous Female Razorbills: How do Females Benefit?" *Behav Ecol Sociobiol* 29 (1992): 455–464; cf. also Richard H. Wagner, "Evidence that Female Razorbills Control Extrapair Copulations," *Behaviour* 118 (1991): 157–169.

38. Joan Roughgarden, Meeko Oishi, and Erol Akçay, "Reproductive Social Behavior: Cooperative Games to Replace Sexual Selection," *Science* 311 (2006): 965–969.

As mentioned before, our *Science* paper provoked mostly emotional outbursts,[39] but for one main exception. A comment by Peter Hurd[40] from the University of Alberta called our attention to the theory of marriage in economics, a theory that we had not been aware of, and that had not, to our knowledge, ever been brought to a biological setting. So, we decided to spend a term discussing and working through examples from the economic theory of marriage initially developed in the 1970s by the economist, Gary Becker,[41] and subsequently extended by others.[42]

Becker's work introduced two ideas, one for the "marriage market," which indicated the payoffs to all possible pairing arrangements, and the other concerned the negotiation between the marriage partners about how to split their family income to satisfy each's preferences. Our group felt some aspects of this theory needed to be tailored considerably before it could be applied to animal families. The payoff to a pair could be understood as the number of young fledged, and the idea of a payoff matrix representing the yield in fledged young from a nest for all possible pairing arrangements seemed okay, needing no modification. But the division of those eggs into those belonging, so to speak, to one or the other partner seemed more problematic. The young from a given pair automatically have an equal number of genes from both parents—there is no way for a nestling to have more genes from one parent than the other and so the parents can't negotiate on how many genes each places in their offspring. On the other hand, it is possible for neighbors to exchange

39. Etta Kavanagh (ed.), "Debating Sexual Selection and Mating Strategies," *Science* 312 (2006): 689–697.

40. Kavanagh, *Debating Sexual Selection*, 692–693.

41. Gary S. Becker, "A Theory of Marriage: Part I," *J Political Economy* 81 (1973): 813–846; Gary S. Becker, "A Theory of Marriage: Part II," *J Political Economy* 82 (1974): S11–S26.

42. Marilyn Manser and Murray Brown, "Marriage and Household Decision-making: A Bargaining Analysis," *Inter Economic Rev* 21 (1980): 31–44; Marjorie B. McElroy and Mary Jean Horney, "Nash-bargained Household Decisions: Toward a Generalization of the Theory of Demand," *Inter Economic Rev* 22 (1981): 333–349; Helen V. Tauchen, Ann Dryden Witte, and Sharon K. Long, "Domestic Violence: A Nonrandom Affair," *Inter Economic Rev* 32 (1991): 491–511; Shelly Lundberg and Robert Pollak, "Separate Spheres Bargaining and the Marriage Market," *J Political Economy* 101 (1993): 988–1010.

parentage, so that some eggs in a nest come from the pair tending the nest, and other eggs from neighbors. Whether such exchanges could be beneficial remained to be investigated, but in any case, such exchanges would be possible, whereas, unlike in human marriage, the family earnings are not divisible between partners. To my knowledge, marriage economic theory doesn't consider exchanges between adjacent married couples, although payments like dowries to place their offspring into marriage could be considered in that framework.

Back when I was writing *Evolution's Rainbow*, I thought the idea of distributed parentage needed to be considered to explain EPPs and EPMs, instead of narratives of cheating, deceit and coercion. The overall idea is one of a "breeding neighborhood." Population genetics has long had the concept of a "genetic neighborhood"[43] to describe the spatial scale according to which alleles share identity by descent. Similarly, there is a spatial scale for the distribution of parentage. One might think of the collection of bird nests in a woods as a system of nests. The nests might be close to one another, and if physically touching one another, would comprise a colony. But the distance between nests could be expanded, as an exploded colony, so to speak, depending on how resources are distributed. It seems to make sense for birds to distribute their eggs into several nests to spread the risk of predation and damage from storms, fire and so forth. Furthermore, by distributing eggs among nests, the adult birds in the area acquire a collective interest in patrolling for predators, and also shared parentage removes an incentive to attack each other's nests. The idea of a breeding neighborhood therefore refers to the spatial scale over which parentage of nests is shared, thereby comprising a system of nests. The nests in an acre of woods can be thought of as a set of overlapping circles representing the breeding neighborhoods of all the nests in the woods.

Erol immediately saw a way to adapt the economic theory of marriage to provide a model for how EPPs could be seen as a system of nests.[44]

43. G. Malécot, *The Mathematics of Heredity* (San Francisco: W. H. Freeman, 1969).

44. Erol Akçay and Joan Roughgarden, "Extra-pair Parentage: A New Theory Based on Transactions in a Cooperative Game," *Evolution Ecol Res* 9 (2007): 1223–1243.

Table 16   Pairing Matrix: Number of Fledglings Produced by a Pair at the
Nest They Would Occupy

| | | *Female Partner* | |
		FEMALE 1	FEMALE 2
Male Partner	MALE 1	4	1
	MALE 2	5	4

Here is how Erol developed his model, a model defined in the behavioral tier, not the evolutionary tier.

One of the first breeding decisions by individual birds is to choose an economic partner. Individuals of different sexes in a breeding neighborhood can be thought of as a pool from which economically monogamous pairs are to be formed. Individuals in such a context can be expected to make adaptive choices that maximize their expected fitness gain for the season. To illustrate, first consider a numerical example using the simplest pairing problem for a breeding neighborhood with two males and two females.

Erol assumed that each possible pair can expect to fledge a certain number of young from the nest they would be occupying, a number that depends on their parental experience as well as ecological features such as territory quality or the nest's exposure to predation. The number of fledglings produced by each pair can then be arranged in a "pairing matrix," as shown in Table 16.

Accordingly, Male 1 and Female 1 can fledge four offspring together, but Male 1 and Female 2 can fledge only one offspring together. Similarly, Male 2 and Female 1 can fledge five young together, whereas Male 2 and Female 2 can fledge four young together.

The fitness payoff to each individual is the total number of genetic offspring produced both on and off their own nests. If individuals are the genetic parents only of offspring in their own nests, the payoffs to each

Table 17 Pairing Matrix: No Extra-Pair Percentage: Fitness Payoffs to Each Member of a Pair for All Possible Pairings

		Female Partner	
		FEMALE 1	FEMALE 2
Male Partner	MALE 1	(4, 4)	(1, 1)
	MALE 2	(5, 5)	(4, 4)

will be simply a repetition of the pairing matrix for each member of the pair as shown in Table 17, where the first entry in each cell is the payoff to the male and the second to the female.

If the pairing takes place to maximize the fitness, Male 2 and Female 1 will pair with each other because each can rear the most young through this collaboration. Meanwhile, Male 1 and Female 2 are stuck with each other, even though each prefers a different partner.

Next, suppose individuals have the option of siring eggs or depositing eggs in other nests. In the present no-extra-pair-parentage arrangement, Male 1 and Female 2 have much to lose if they pair with each other, relative to what they could produce with Female 1 and Male 2 respectively. Therefore, Male 1 has an incentive to negotiate with Male 2 to let him pair with Female 1. Male 1 can offer Male 2 access to Female 1 such that Male 2 fertilizes, say, two of the eggs in the clutch that Male 1 would raise with Female 1. Such an arrangement would give Male 2 a total of six offspring, four from his clutch with Female 2, and two from the nest of Male 1 and Female 1. Conversely, Male 1 would sire two offspring in his clutch with Female 1, leading to the following arrangement of payoffs in Table 18.

This "deal" would be acceptable to both males, since both would be siring more offspring relative to the arrangement with no extra-pair paternity. In this way, extra-pair paternity can arise as a side-payment that stabilizes a certain economic pairing arrangement. Incidentally, Male 1 does not need to guarantee Male 2 the paternity of exactly two of

Table 18   Payoff Matrix: With Extra-Pair Percentage: Fitness Payoffs to Each
Member of a Pair for All Possible Pairings

| | | *Female Partner* | |
		FEMALE 1	FEMALE 2
Male Partner	MALE 1	(2, 4)	(1, 1)
	MALE 2	(5, 5)	(6, 4)

the eggs in his nest. Rather, Male 2 simply requires that the expected
number of offspring he sires in Male 1's nest be two, which he can ensure
by having adequate copulation access to Female 1.

What about the females? In the above arrangement, Female 1 seems to
be losing out, since she pairs with Male 1 and produces four offspring instead of five. Two things might happen in this case. First, Female 1 can be
compensated for this loss with a similar agreement between the females:
to secure Female 1's cooperation, Female 2 can allow her to deposit two
eggs (fertilized by Male 2) into the nest of Female 2 with Male 2. This
would bring the payoff to Female 1 to six, while keeping the payoff to
Female 2 at two, which is still greater than the one offspring she would
mother if she paired with Male 1 instead. This outcome predicts extra-
pair maternity in addition to extra-pair paternity.

Alternatively, Female 1 might not be compensated and thus would
have no incentive to cooperate on this arrangement. She might, for example, refuse to copulate with Male 2, leading to forced extra-pair copulations or other forms of conflict behavior between Female 1 and the
males. The presence and extent of such conflict would be determined by
the outside options available to the female as well as other benefits that
might be involved. Such sexual conflict would not be fundamental to a
male-female relationship, but a special condition derived from a failure
at negotiation, as can occur in any labor dispute regardless of the sexes of
the parties involved.

Erol has mathematically analyzed the pairing-payoff matrices, such as in Table 17, to determine the general conditions under which EPP and EPM are expected as side-payments to stabilize a pairing arrangement. The analysis treats males and females separately with two separate two-player games, one between adjacent males to determine the EPPs and one between adjacent females to determine the EPMs. When a side-payment is warranted, the size of the payment can be computed as a Nash bargaining solution in the games between neighbors of the same sex.

Erol demonstrated how a theory for extra-pair parentage based on the intuition of side-payments can account for the known regularities in how EPPs are distributed. For example, side-payment model predicts a negative relationship between paternal care and levels of extra-pair paternity. The relative importance of capable motherhood becomes greater when males contribute little themselves to offspring rearing. Therefore, a male is willing to allow more extra-pair paternity to have the opportunity to pair with a female who is a successful mother if he is not going to provide any care himself.

More generally, the condition for extra-pair parentage to occur is where high-productivity individuals are induced to pair with partners of lower productivity whom they would not otherwise prefer. Such high-productivity individuals need to be compensated for the loss of production they would have attained with their preferred partner to induce their pairing with a lower-productivity partner. At the same time, lower productivity individuals must be willing to concede some proportion of their nests' parentage in exchange for the opportunity to pair with high-productivity individuals. Therefore, high-productivity individuals will on average have higher proportions of EPPs because they will be receiving payments from lower-productivity individuals, resulting in a positive correlation between individual productivity and the amount of extra-pair paternity.

In summary, sexual-selection theory interprets the higher proportion of EPPs accrued by some individuals as reflecting female choice for males with good genes. Sexual-conflict theory interprets high EPPs as resulting when males do not prevent female infidelity, that is, if males contribute no parental care to begin with, they have nothing to withhold as punishment for infidelity. In contrast, social-selection theory interprets the higher EPPs accrued by some males as accumulated side-payments from other males paying to be paired with females whom they could not otherwise pair with.

The side-payment model is also relevant to patterns of extra-pair maternity. The most consistent correlate of extra-pair maternity is annual fecundity.[45] Extra-pair maternity as a percentage of total offspring increases with increasing fecundity. Because maternity comes in strictly discrete units (i.e., eggs), unlike paternity which is probabilistic and thus can vary more or less continuously (in terms of its expected value), a maternity exchange may not be feasible if it requires the transfer of, say, half an egg. In species with high fecundity, however, transactions would be prohibited less often by the discrete nature of maternity, because the proportion of maternity in a clutch approaches a continuous variable. Therefore, all else being equal, high-fecundity species should exhibit more extra-pair maternity.

The side-payment model for EPPs presented here is similar in spirit to a model proposed by Shellman-Reeve and Reeve[46] who assumed that the social mates receive paternity in exchange for the parental care they supply. Their model assumed that females prefer to mate with an extra-pair male, because of some genetic benefit, but at the same time need full-time parental provisioning from their social mate. The side-payment approach presented here shows that genetic benefits do not need to be assumed to underlie the transactions.

More broadly, the side-payment model extends the reproductive transactions theory for cooperative breeding groups by Sandy Vehrencamp[47] to exchanges between monogamous pairs. Cooperative game

45. E. Geffen and Y. Yom-Tov, "Factors Affecting the Rates of Intraspecific Nest Parasitism Among Anseriformes and Galliformes," *Anim Behav* 62 (2001): 1027–1038; K. E. Arnold and I. P. F. Owens, "Extra-pair Paternity and Egg-dumping in Birds: Life History, Parental Care and the Risk of Retaliation," *Proc R. Soc Lond B* 296 (2002): 1263–1269.

46. J. S. Shellman-Reeve and H. K. Reeve, "Extra-pair Paternity as the Result of Reproductive Transactions Between Paired Mates," *Proc R Soc Lond B* 267 (2000): 2543–2546.

47. S. Vehrencamp, "A Model for the Evolution of Despotic Versus Egalitarian Societies," *Anim. Behav* 31: 667–682.

theory improves upon reproductive transactions theory by allowing the coalition structure or group composition as well as the pattern of payoff distribution or skew to emerge from the model. Here, the groups are monogamous pairs but the approach is extensible to polygynous and polyandrous groups as well. Moreover, parentage may be viewed as the currency that can buy different commodities and services. These include provisioning, nest defence, and risk sharing through dispersing parentage over different nests.

I'd like to end this chapter with some theoretical "to-do" issues. It's ironic that the explanation of many of the phenomena that motivated my inquiry into replacing sexual-selection theory have had to be postponed for so long. Here are some high priority items:

First is the extension of the family-dynamics theory of this and the preceding chapter to include more economic roles than merely a "male" and "female." Many species contain multiple types of individuals within each sex, multiple genders per sex, including pre-zygotic and post-zygotic helpers. What ecological situations support family structures containing these multiple economic roles or social niches?

Second is to develop a theory for when sex-role reversal takes place. Just why has the male/female negotiation over the division of labor led to the males having the higher parental investment, and females the lower investment, in species ranging from pipefish to birds like the jacana and phallarope?

Third is to account for the division of sex in coalition structures—do males hang out together, and females together, or are social groups mixed with respect to sex? What are the conditions for inclusion in social groups? This question of who gets to belong in a social group returns us to the original motivation for sexual-selection theory—the peacock's tail. I have previously suggested that secondary sexual characters like the peacock's tail are expensive "admission tickets" to monopolistic power-holding cliques. The peacock's tail may not matter much to a female. But a male lacking expensive ornamentation would suffer discriminatory exclusion. Our lab plans to develop this idea further in the near future.

To conclude this book, what can we now say about whether biological nature is really selfish?

Social versus Sexual Selection

Since the origin of evolutionary biology, Darwinism has been synonymous with competition and selfishness. Evolution is identified with the phrase, "survival of the fittest" that was coined by Herbert Spencer, not by Charles Darwin himself. Darwin hypothesized that evolution occurs through descent with modification from common ancestors. Because this hypothesis is now established as fact, it would seem that science has also determined that biological nature consists of a competitive, selfish, struggle for survival. But the 200 years since Darwin's birthday have confirmed solely Darwin's hypothesis about descent and common ancestry—whereas confirming Spencer's interpretation of evolution is another matter altogether.

This book does not dwell on disturbing theological or ethical implications of the premise that biological nature is "red, in tooth and claw," to

quote Alfred, Lord Tennyson's poem of 1850. Nor does this book histori-
cally deconstruct Darwin or Spencer by situating them in their Victorian
culture, nor does it provide a feminist-theoretic critique of the wealth
and prestige attending those privileged to define the criteria for nor-
malcy. And this book does not declaim social-Darwinist political theories
or agendas. This book is different. This book invites debate about
whether Spencer and his ideological descendants today correctly describe and explain biological nature. This book focusses on the scientific
accuracy of claims that evolution rests on competition, selfishness, and
conflict. One may find metaphors like "the selfish gene" and "the battle
of the sexes" appealing or repugnant. Regardless of what we would like
the truth to be, the issue before us is whether such metaphors correctly
characterize biological nature.

It's fair to say that most social behavior relates in some way to reproduction—mating and the rearing of offspring. Investigating how reproductive social behavior evolves offers perhaps the best test case of
whether evolution really does rest on competition and reward the selfish. The evolutionary theory specifically for reproductive social behavior
is called "sexual-selection theory," and it portrays both the process and
outcome of evolution as selfish, deceitful, and coercive. I have come to
doubt that sexual-selection theory is correct.

Since Darwin initiated the topic of sexual selection with a narrow purpose of explaining exaggerated male ornaments, such as the peacock's
tail, it has accumulated propositions and corollaries that collectively
comprise a "system." An explanatory system such as the sexual-selection
system, offers a continuous narrative to account for a large suite of phenomena, in this instance, phenomena ranging from why sex, male, and
female exist to begin with, extending through the characteristics of sexual reproduction, such as the organization and dynamics of animal family life, and culminating in claims about the essential nature of humans.
I term this extended narrative the "sexual-selection system" for the evolution of sex, gender, and sexuality. It tells a story of selfishness, deceit,
and coercion as naturally inevitable. This is the narrative that I have come
to think is incorrect. To determine whether it is indeed incorrect, one must
test it. The first step in a scientific test is to formulate an alternative that

Table 19   Contrast between Evolutionary Systems of Sex, Gender, and Sexuality

Social Selection	*Sexual Selection*
1. Sex to balance genetic portfolio.	Sex to eliminate deleterious mutations.
2. Sperm/egg binary to maximize gametic contact rate.	Sperm/egg binary result of gametic sexual conflict.
3. Both sexes in single body are primitive.	Single sexes in separate bodies are primitive.
4. Sex roles are negotiated.	Male/female is passionate/coy.
5. Social behavior as offspring producing system.	Social behavior as mating system.
6. Female choice maximizes offspring quantity.	Female choice maximizes offspring genetic quality.
7. Males equivalent in genetic quality.	Males form hierarchy of genetic quality.
8. Bateman's principle invalid.	Bateman's principle foundational.
9. Behavior as development.	Behavior as evolution.
10. Social outcome as NBS or NCE.	Social outcome as ESS.
11. Male/female investment equal, egg ≈ total ejaculate.	Male/female investment unequal, egg ≫ individual sperm.
12. Sexual cooperation primitive, sexual conflict incidental.	Sexual conflict primitive, sexual cooperation illusionary.
13. Male promiscuity as tactic of last resort.	Male promiscuity to increase paternity.
14. Monogamy as efficient team for offspring production.	Monogamy as female entrapment or males lacking alternatives.
15. Extra-pair paternity and egg transfers as side-payments to form pairing arrangements.	Extra-pair paternity as female infidelity escaping male punishment, and egg dumping as brood parasitism.
16. Peacock tails as admission tickets to powerful cliques.	Peacock tails as ornaments advertising desirable genes.
17. Monomorphism indicates absence of cliques.	Monomorphism reflects no female aesthetic for ornaments.
18. Sex-role reversal from male/female labor negotiation.	Sex-role reversal from reversed operational sex ratio.
19. Gender multiplicity reflects pre-zygotic and post-zygotic helpers, social niches.	Multiple male morphs as alternative mating strategies, sexual parasites, sneakers, who steal investment.

Continued

Table 19 (continued)

Social Selection	Sexual Selection
20. Feminine males expressing body English to serve as marriage brokers.	Feminine males as female mimics to deceive masculine males to steal their matings.
21. Masculine females with body English for territoriality.	Masculine females with imperfect suppression of male traits.
22. Homosexuality is physical intimacy to coordinate action and sense team welfare.	Homosexuality is accidental, deceptive, or a genetic disease maintained by sexually antagonistic pleiotropy.
23. Human attractiveness responds to health and behavioral compatibility to raise young.	Human attractiveness in males signals good genes and in females signals fecundity.
24. Human brain necessary for social infrastructure in which offspring are reared.	Human brain in males as an ornament to attract females and in females an imperfectly expressed male trait.
25. Human rape as social power.	Human rape as reproductive strategy.
26. Human EPP as distributed parentage and alliance.	Human EPP as genetic upgrade expressed when punishment is limited.

offers a substantially different explanation for the same set of phenomena that the reigning theory purports to explain. Then in time, new information may settle the matter.

In Chapters 3 to 10 I have summarized the research of my laboratory during the last five years to develop an alternative system for the evolution of sex, gender and sexuality, a system I term the "social-selection system." Table 19 presents a list that contrasts 26 elements of both the social-selection and sexual-selection systems, point by point. The initial part of Table 19 pertains to the 15 foundational elements discussed in Chapters 3 to 10, and the final part offers 11 more items that are introduced here.

Here then is a point-by-point summary of the differences between social selection and sexual selection as alternative theoretical systems to account for the evolution of sex, gender and sexuality.

1. Origin of Sexual Reproduction. According to sexual selection, sexual reproduction evolved from asexual reproduction as a mechanism to cleanse the gene pool of deleterious mutations. According to social selection, sexual reproduction evolved from asexual reproduction to maintain a diverse gene pool needed for long-term population-survival in an ever-fluctuating environment.

2. Origin of Sperm/Egg Binary. According to sexual selection, egg and sperm result from primordial sexual conflict—a battle of the gametes. According to social selection, sperm and egg maximize the rate of gametic contacts that produce viable zygotes.

3. Origin of Male/Female Binary. According to sexual selection, the whole-organism binary is taken at a starting point, and hermaphrodism is viewed as a special case arising when population density is extremely low or when it is better to be one sex when small and another when large. According to social selection, hermaphrodism is the starting point and separate sexes reflect a specialization for the "home delivery" of sperm.

4. Sex Roles. According to sexual selection, males and females conform to the near-universal templates that males are "passionate" and females "coy." According to social selection, what each sex does is subject to negotiation in local circumstances and statistical regularities in sex roles reflect commonness of circumstance.

5. Purpose of Social Behavior. According to sexual selection, reproductive social behavior comprises a "mating system." Within a mating system, natural selection arises from differences in "mating success" and particular behaviors are viewed as yielding mating opportunities. The females are regarded as a "limiting resource" for males, and males compete for access and control of mating opportunities with females. According to social selection, reproductive social behavior comprises an "offspring producing system." Natural selection arises from differences in number of offspring successfully reared, and particular behaviors are viewed as contributing to producing offspring and to building or maintaining the social infrastructure within which offspring are reared. The male/female social dynamic determines bargains and exchanges side-payments to control offspring and manage the social infrastructure.

6. Objective of Female Mate Choice. According to sexual selection, females select mates for their genetic quality to endow their own sons with the traits they found attractive, an indirect benefit of

mate choice. According to social selection, a female chooses mates to maximize the number of young she can produce that are successfully reared by her own efforts plus help from her mates and plus assistance from the social infrastructure. She chooses based on the direct benefits promised by a male discounted by the probability that the male will renege on or be prevented from delivering on his promise. A premium will be placed on the compatibility and health of the prospective partner. Health is important not as an indicator of genetic quality but as a sign of competency to deliver on promised direct benefits.

7.   Organization of Male Genetic Quality. According to sexual selection, males can be ranked in a hierarchy of genetic quality. According to social selection, no hierarchy of genetic quality among males exists. All males are equivalent in genetic quality, except a rare fraction that obviously contain deleterious mutations and are present in a mutation-selection balance.

8.   Bateman's Principle. In 1948, Bateman reported that for males "fertility is seldom likely to be limited by sperm production but rather by the number of inseminations or the number of females available to him." Similarly, he claimed to have found in his flies an "undiscriminating eagerness in males and discriminating passivity in females" in accord with the sexual-selection narrative. The Bateman experiments are widely cited in papers and textbooks as foundational to sexual selection. According to social selection, Bateman's conclusions are suspect and discredited.

9.   Causes of Behavior. Sexual selection views behavior as caused by genetic processes occurring on a between-generation time scale, like morphological traits. Social selection views behavior as caused by developmental processes on a within-generation timescale reflecting accumulated experience. Both sexual selection and social selection agree that behavior ultimately results from evolution, but sexual selection views evolution as directly causing social behavior, whereas social selection views social behavior as developmentally emergent from evolved payoffs.

10. Social Outcome. Sexual selection views social outcomes as an evolutionarily stable strategy (ESS)—a situation for both players such that a mutant allele for any other strategy cannot increase when rare. This single genetic tier requires seeking a genetic basis to behavior. Social selection views behavior as resulting from dynamics

occurring in two tiers—development nested within evolution. The developmental dynamics employ both cooperative and competitive game theory resulting in either a Nash bargaining solution (NBS) or a Nash competitive equilibrium (NCE). Cooperative solutions may be attained by the parties playing as a team with coordinated tactics and with the perception of shared goals made possible through intimate friendship. Although such teamwork involves cooperation among teammates, it may involve disagreement on how to distribute team earnings, and the exclusion of nonteammates through prejudice. Social selection also envisions an evolutionary tier in which the game-theoretic payoff matrices and rules of play evolve based on traditional population-genetic dynamics.

11. Parental Investment. According to sexual selection, the female has a higher parental investment than the male because the egg is bigger than the sperm. According to social selection, male and female parental investments are more or less the same initially. An ejaculate might typically contain 10^6 sperm, whereas an egg is typically 10^6 times as large as a sperm. So the size of the ejaculate and egg are often about the same order of magnitude.

12. Sexual Conflict. According to sexual selection, a male and female are fundamentally in conflict and their cooperation is a possible (and unlikely) secondary development. According to social selection, male and female mates begin with a cooperative relationship because they have committed themselves to a common "bank account" of evolutionary success. Their offspring represent indivisible earnings from a common investment. As such, conflict develops only secondarily if a division of labor cannot be successfully negotiated.

13. Male Promiscuity. According to sexual selection, males are naturally and universally promiscuous, reflecting the low parental investment of a sperm compared to an egg. According to social selection, male promiscuity is a strategy of last resort that occurs when males are excluded from control of offspring rearing.

14. Monogamy. In sexual-selection theory, monogamy is entrapment of males by females, or a default when no other mates are available. Social selection distinguishes two forms of monogamy: economic monogamy—rearing offspring in teams of one male with one female, and genetic monogamy—mating solely between one male and one female. Economic monogamy does not coincide with

genetic monogamy when distributed parentage is adaptive. Economic monogamy emerges in situations where offspring rearing is most efficiently done in male-female teams, rather than as solitary individuals, or in teams with two individuals of the same sex, or in teams of more than two individuals. Genetic monogamy emerges when distributed parentage is nonadvantageous.

15.   Extra-Pair Parentage. Sexual selection views extra-pair paternity as reflecting females seeking genetic benefits by mating with an extra-pair male while taking advantage of the pair male's parental care. EPPs are prevented when males can punish infidelity. Females who dump eggs in adjacent nests are intra-specific brood parasites who deceive nearby nest-pairs, tricking them into raising her young. According to social selection, extra-pair parentage reflects a system of genetic side payments that stabilizes an economic pairing arrangement among individuals with asymmetrical capacities for offspring rearing. Distributed parentage also spreads the risk of nest mortality across a network of nests, acting as a social insurance policy.

The items above were discussed extensively in the preceding chapters. Next I highlight items that previously were mentioned briefly or not at all.

16.   Secondary Sexual Characters. According to sexual selection, secondary sexual characters like the peacock's tail and the stag's antlers are favored by females in mate choice so that their own sons will be attractive and victorious in the next generation. The "beauty" a female perceives in a male's ornaments is how she apprehends a male's good genes. Male ornaments are "condition indicators" of genetic quality. According to social selection, ornaments, both male and female, serve as "admission tickets" to power-holding cliques that control the opportunity for successful rearing of offspring. Admission tickets are expensive because the advantage to membership in a clique resides in the power of monopoly, which is diluted when membership is expanded. By requiring a high price of admission, the monopolistic coalition is kept exclusive, maximizing benefit to those within. Ornamental admission tickets belong to a class of traits called "social-inclusionary traits" that are needed to participate in the social infrastructure within which offspring are reared. Not possessing such traits, or

not participating in social-inclusionary behaviors, is reproductively lethal.

17. Sexual Monomorphism. According to sexual selection, male appearance is indistinguishable from female appearance when females happen to lack a sense of aesthetic. In social-selection theory, sexual monomorphism reflects the absence of same-sex power-holding cliques whose membership requires admission tickets. This should occur in ecological situations where the economically efficient coalition is the coalition of the whole.

18. Sex-Role Reversal. According to sexual selection, reversal occurs when, for some unknown reason, the male happens to provide more parental investment than the female, which then causes males to be in short supply for mating relative to females. This imbalance in supply causes females to compete with one another for access to males, resulting in the females evolving to be the showy sex while the male remains drab. However, this account does explain sex-role reversal. Sexual selection fails to say why the male should happen to be the sex providing the higher parental investment in such species, contradicting its own premise that males should be providing the lower parental investment because the sperm is smaller than the egg. According to social selection, reversed sex roles are not problematic, because sex roles are negotiated in local ecological situations anyway. A male's interest is furthered by securing some control of the eggs and, thereby, retains at least partial control of his evolutionary destiny. In securing control of the eggs, in some ecological circumstances, he might incidentally wind up with a higher parental investment than the female, setting the stage for a reversal in the supply of females relative to males.

19. Gender Multiplicity. Many species have multiple templates of males and females. I term each template a "gender." According to sexual selection, the large territory-holding male gender is taken as the reference male, and the other genders of males are considered as "alternative mating strategies" and regarded as "sexual parasites." According to social selection, economic theory for the elemental one-male-one-female economic team is extended to larger teams with more "social niches." Some team members are "pre-zygotic helpers"—animals who assist in bringing about courtship and mating, together with "post-zygotic helpers"— members who remain at the nest to help rear the offspring that

have already been born. Those not included in the reproductive social group's coalition form other arrangements to oppose it, either singly or in coalitions of their own. Thus, multiple male templates are not alternative mating strategies but multiple social niches within reproductive social groups.

20. Feminine Males. According to sexual selection, feminine males are termed "female mimics"—sexual parasites who steal the reproductive investment of a territory-holding males through deceit. A female mimic is disguised as a female to enter the territory-holding male's harem and mate with his females. According to social selection, markings and colors on animals represent "body English"—how animals tell one another what their social role is, what their intentions are, and what activities they promise to perform. Feminine males are participating in a conversation on topics and with words used more frequently by females than by masculine males.

21. Masculine Females. In sexual selection, masculine females are discussed as "female ornaments"—hanging skin flaps (wattles), colored feathers, antlers, etc. usually considered as male ornaments. Darwin dismissed out-of-place ornaments as male traits expressed in females as a developmental error. According to social selection, masculine females are the reverse of feminine males, namely, a female using body English to converse on topics and with words used more frequently by males than by the feminine females. Such conversations might involve establishing and defending territories in species where this task is sometimes carried out by females, but usually by males.

22. Homosexuality. According to sexual selection, homosexuality is an inadvertent mistake, a deception, or a deleterious trait maintained through peculiar inheritance. Homosexuality occurs when, for example, a small male sneaks into the territory of a large male, allows him to tire by acquiescing to homosexual copulation, and then mates with the females in his harem. Or homosexuality is a disease caused by genes that decrease fitness in one sex, but increase fitness in the other sex (sexually antagonistic pleiotropy).[1] According to social selection, homosexuality is natural and adaptive to all

1. S. Gavrilets and W.R. Rice, "Genetic Models of Homosexuality: Generating Testable Predictions," *Proceed Royal Society London B* 273 (2007): 3031–3038.

participants and both sexes. Homosexuality along with mutual grooming, preening, sleeping, tongue rubbing, and interlocking vocalizations allows animals to work together as a team—to coordinate actions and tactilely to sense one another's welfare.

The final four items pertain to how evolutionary psychologists extrapolate sexual selection directly to human behavior, contrasted with social selection's alternative explanation.

23. Human Attractiveness. According to sexual selection, women are supposed to find men handsome who display traits indicating their genetic quality. Men are supposed to be naturally promiscuous.[2] According to social selection, males and females choose each other equally with the criterion for both being compatibility of circumstance, temperament, and inclination that underlies effectiveness at raising offspring in the context of a human social infrastructure.

24. Human Brain. According to sexual selection, the human brain is a secondary sexual character, the counterpart of the peacock's tail, an ornament used by men to attract women.[3] The problem then is to explain why women have brains. According to sexual selection, a woman's brain is a "female ornament," seemingly as out of place in a woman as a gaudy tail on a peahen. Sexual selection then postulates that females need big brains to appreciate the brains of men. In contrast, social selection views the human brain as a trait needed to participate in the social infrastructure within which offspring are reared and is equally necessary in both men and women for successful offspring production.

25. Human Rape. Sexual selection views human rape as an evolutionary strategy whereby men who are not preferred as mates by women manage to reproduce through coercion.[4] Social selection views human rape as domination, an exercise of power that can be

---

2. D. Buss, *The Evolution of Desire* (New York: Basic Books, 1994).

3. G. Miller, *The Mating Mind, How Sexual Choice Shaped the Evolution of Human Nature* (New York: Anchor Books, 2000).

4. R. Thornhill and C. Palmer, *A Natural History of Rape: Biological Bases of Sexual Coercion* (Cambridge: MIT Press, 2000); cf. critiques in: C.B. Travis, ed., *Evolution, Gender, and Rape. A Bradford Book* (Cambridge: MIT Press, 2003); J. Roughgarden, review of *Evolution, Gender and Rape* by J. Travis, ed., *Ethology* 110 (2004): 76–78.

either homosexual or heterosexual, and that has no direct relation to reproduction.

26. Human Extra-Pair Paternity (EPP). Sexual selection views EPP as women soliciting mating from men who are genetically superior to their husbands, provided they can retain their husband's parental investment.[5] Social selection views EPP as distributed parentage to spread the risk of offspring rearing, avoiding the hazards of having "all eggs in one basket," and procuring a distributed commitment to protection, safety, and resource sharing through political alliance.

Tables 19 reveals that the theory of sexual selection has now become a theoretical system of thought, a stream of logically interconnected propositions that amount to a philosophy of biological nature. Social selection then stands as an alternative theoretical system.

Sexual-selection advocates resist seeing the whole of sexual selection as a theoretical system and prefer to view it as a collection of disarticulated propositions each of which might be individually true of false without repercussions amongst them all. As Gowaty writes, "Roughgarden says that sexual selection is a *system*; most investigators think sexual selection is *an hypothesis* to explain the evolution of bizarre or showy traits."[6] Yet, as this table reveals, sexual selection has obviously expanded far beyond its early preoccupation with the peacock's tail, and the logical and empirical interdependencies among the various propositions are evident.

Many have suggested that a synthesis of social selection and sexual selection must surely be possible, given that both are evolutionary theories. Here's the rub. I think that sexual selection is incorrect on each of the 26 propositions in Table 19. So, from my vantage point, I cannot discern how social selection would be strengthened by adding one of the elements of sexual selection.

5. S.W. Gangestad and R. Thornhill, "Human Oestrus," *Proc R Soc B* 275 (2008): 991–1000.

6. Patricia Adair Gowaty, "Perception Bias, Social Inclusion, and Sexual Selection: Power Dynamics in Science and Nature," in *Controversies in Science & Technology. Volume 2: From Climate to Chromosomes*, ed. Daniel Lee Kleinman et al. (New Rochelle: Mary Ann Liebert, 2008), 401–420.

Indeed, consider the implications of sexual selection being incorrect on all 26 elements in the table. Could this be an unfortunate coincidence? As a back-of-the-envelope calculation, suppose, conservatively, that the chance of sexual selection being incorrect on any one element is ½. Actually, sexual-selection advocates think that the chance of sexual selection being wrong on any of the 26 points is near zero. Nonetheless, if the probability of sexual selection being wrong on any one element is as high as ½, then the odds of being incorrect on every element assuming independence is $(½)^{26}$, which equals about 1.5×10^{-8}. This low probability is exceedingly significant statistically and implies that the assumption of independence must be rejected. It implies instead that the 26 elements are coupled. Although one may tinker with this calculation, the point is that some feature common to all 26 propositions in sexual selection must exist to explain why they are all incorrect at the same time. That feature is that all 26 points derive from a common view of natural behavior predicated on selfishness, deception, and genetic weeding. If this view of biological nature is wrong, then deriving any 27th or 28th additional element for the sexual-selection system will fail as well. Thus, the sexual-system cannot be somehow repaired or sanitized. Its foundation is incorrect.

In contrast, the reason social selection seems to be correct across all 26 issues listed in the table is that the elements of social selection derive from an alternative view of biological nature, a view predicated on teamwork, honesty, and genetic equality. The fact that conflict and deception certainly occur in nature, too, does not alter the possible truth of this alternative view because conflict and deception can be understood as secondary imperfections of processes that promote cooperative behavioral and evolutionary outcomes.

The broad failure of sexual selection is not a threat to evolutionary biology. To the contrary, rejecting hypotheses is the way science works, and new hypotheses are suggested in their place. A lot has been learned about the natural history of animal behavior during attempts to confirm sexual selection. That knowledge doesn't disappear because sexual selection turns out to be incorrect. Instead, that information is more compatible with social selection. On issues ranging from why sex exists, to why gametes come in two sizes, to how gamete production is packaged in bodies, to the role of experience rather than genes in determining behavior, to how birds help each other at nests, to how offspring rearing is distributed among neighbors, to how animals share physical intimacy, to how variable sex roles are, and on and on—the huge panoply of facts about animal life discovered in recent decades, all accord nicely with social selection, much more so than with sexual selection.

As often stated in this book, the issue before us is not whether a biological nature predicated on selfishness, deception, and genetic hierarchy is appealing or repugnant compared with a biological nature predicated on teamwork, honesty, and genetic equality. The issue is which of these views of biological nature is true. I believe I have shown that the overwhelming weight of data and theory reveal that the selfish-gene picture does not truly and accurately describe biological nature.

Some feel that the selfish gene metaphor does no more than express a darkly poetic vision of natural dystopia, an entertaining hyperbole for those who enjoy black humor or tragic plays, but otherwise without consequence. I believe I have shown that the selfish gene metaphor is more than "mere poetry." It underwrites an extensive scientific theory that purports to explain some of the basic questions of life that humankind has posed since the dawn of recorded history. Yet, because that metaphor is inaccurate, a huge body of scientific theory is incorrect. I believe I have sketched a new scientific theory, social selection, which addresses the basic questions of life that sexual-selection theory addressed, but does so with a picture that presently appears true and accurate. Time will tell whether social selection is indeed correct or whether some substantial modification or third approach is needed.

I hope that discussion of our own biological nature in the humanities and social sciences, and throughout the public generally, will engage the scientific accuracy of philosophical outlooks claimed to follow from science, rather than accept such claims as given truths. There's still much to be done in the basic science of biology apart from applications to medicine, technology or policy, and the welcoming door is open. Come on in.

Index

Indexer:	Live Oaks Indexing
Composition:	Aptara, Inc.
Text:	10/14 Palatino
Display:	Univers Condensed Light 47, Bauer Bodoni
Printer and Binder:	Maple-Vail Book Mfg. Group